SYSTEM LEVEL DESIGN OF
RECONFIGURABLE SYSTEMS-ON-CHIP

T0143026

System Level Design of Reconfigurable Systems-on-Chip

Edited by

NIKOLAOS S. VOROS
INTRACOM S.A., Patra, Greece

and

KONSTANTINOS MASSELOS
Imperial College of Science Technology and Medicine, London, U.K.

 Springer

A C.I.P. Catalogue record for this book is available from the Library of Congress.

ISBN-13 978-1-4419-3864-0 (PB)
ISBN-13 978-0-387-26104-1 (e-book)

Published by Springer,
P.O. Box 17, 3300 AA Dordrecht, The Netherlands.

www.springeronline.com

Printed on acid-free paper

Contents

Contributing Authors

Miroslav Cupak,
IMEC, Kapeldreef 75, B-3001 Leuven, Belgium

Konstantinos Masselos
Imperial College of Science Technology and Medicine, Exhibition Road, London, SW7 2BT, United Kingdom

Marko Pettissalo
Nokia Technology Platforms, P.O.Box 50, FIN-90571 Oulu, Finland

Yang Qu
VTT Electronics, P.O.Box 1100, FIN-90571 Oulu, Finland

Luc Rijnders
IMEC, Kapeldreef 75, B-3001 Leuven, Belgium

Kari Tiensyrjä
VTT Electronics, P.O.Box 1100, FIN-90571 Oulu, Finland

Nikolaos S. Voros
INTRACOM S.A., 254 Panepistimiou str., 26443, Patra, Greece

Preface

This book presents the perspective of the ADRIATIC project for the design of reconfigurable systems-on-chip, as perceived in the course of the research during 2001 - 2004. The project provided: (a) a high-level hardware/software co-design and co-verification methodology and tools for reconfigurable systems-on-chip, supplemented with back-end design tools for the implementation of the reconfigurable logic blocks of the chip, (b) the definition of the technological requirements for reconfigurable processors for wireless terminals and (c) the implementation of MPEG-4, WCDMA and WLAN design cases to validate the methodology and tools.

Reconfigurability is becoming an important part of System-on-Chip (SoC) design to cope with the increasing demands for simultaneous flexibility and computational power. Current hardware/software co-design methodologies provide little support for dealing with the additional design dimension introduced. Further support at the system-level is needed for the identification and modelling of dynamically re-configurable function blocks, for efficient design space exploration, partitioning and mapping, and for performance evaluation. The overhead effects, e.g. context switching and configuration data, should be included in the modelling already at the system-level in order to produce credible information for decision-making.

This book focuses on hardware/software co-design applied for reconfigurable SoCs. We discuss exploration of additional requirements due to reconfigurability, report extensions to two C++ based languages/methodologies, SystemC and OCAPI-XL, to support those requirements, and present results of three case studies in the wireless and multimedia communication domain that were used for the validation of the approaches.

The book includes nine chapters, divided in three parts: Part A contains Chapters 1 – 3 and provides an introduction to reconfigurable systems-on-chip; Part B contains Chapters 4 – 6 and describes in detail the proposed system level design methodology and the associated tools; Part C, which contains Chapters 7 – 9, provides the details of applying the proposed methodology in practice.

Acknowledgments

The research work that provided the material for this book was carried out during 2001 – 2004 mainly in the ADRIATIC Project (Advanced Methodology for Designing Reconfigurable SoC and Application-Targeted IP-entities in wireless Communications) supported partially by the European Commission under the contract IST-2000-30049. Guidance and comments of Mr Ronan Burgess, Dr Lech Jozwiak and Dr Mark Hellyar on research direction are highly appreciated.

In addition to the authors, the contributions of the following project members and partners' personnel are gratefully acknowledged: Antti Anttonen, Spyros Blionas, Kristof Denolf, Klaus Kronlöf, Tarja Leinonen, Dimitris Metafas, Robert Pasko, Antti Pelkonen, Konstantinos Potamianos, Tapio Rautio, Geert Vanmeerbeeck, Serge Vernalde, Peter Vos, Erik Watzeels, Matti Weisssenfelt and Yan Zhang.

Of them, the editors express their special thanks to Antti Pelkonen and Yan Zhang for their valuable contributions to Chapter 5 and Chapter 9, Robert Pasko and Geert Vanmeerbeeck for their valuable contributions to Chapter 6, Kristof Denolf and Peter Vos for their substantial contributions to Chapter 7 and Serge Vernalde and Erik Watzeels for management related issues.

PART A

RECONFIGURABLE SYSTEMS

Chapter 1

INTRODUCTION TO RECONFIGURABLE HARDWARE

Konstantinos Masselos[1,2] and Nikolaos S. Voros[1]
[1] *INTRACOM S.A., Hellenic Telecommunications and Electronics Industry, Greece*
[2] *Currently with Imperial College of Science Technology and Medicine, United Kingdom*

Abstract: This chapter introduces the reader to main concepts of reconfigurable computing and reconfigurable hardware. Different types of reconfiguration are discussed. A detailed classification of reconfigurable architectures with respect to the granularity of their building blocks, the reconfiguration scheme and the system level coupling is also presented.

Key words: Reconfigurable hardware, reconfigurable architectures, reconfiguration, reconfigurable computing

1. RECONFIGURABLE COMPUTING AND RECONFIGURABLE HARDWARE

Reconfigurable computing refers to systems incorporating some form of hardware programmability–customizing how the hardware is used using a number of physical control points [2]. These control points can then be changed periodically in order to execute different applications using the same hardware. Reconfigurable hardware offers a good balance between implementation efficiency and flexibility as shown in Figure 1-1. This is because reconfigurable hardware combines post-fabrication programmability with the spatial (parallel) computation style [2] of application specific integrated circuits (ASICs), which is more efficient in comparison to the temporal (sequential) computation style of instruction set processors.

Due to the increasing flexibility requirements (e.g. for adaptation to different evolving standards and operating conditions) that are imposed by computationally intensive applications such as wireless communications,

15

N.S. Voros and K. Masselos (eds.), System Level Design of Reconfigurable Systems-on-Chips, 15-26.
© 2005 *Springer. Printed in the Netherlands.*

devices need to be highly adaptable to the running applications. On the other hand, efficient realizations of such applications are required, especially in the resources they use during deployment, where power consumption must be traded against perceived quality of the application. The contradictory requirements for flexibility and implementation efficiency cannot be satisfied by conventional instruction set processors and ASICs. Reconfigurable hardware forms an interesting implementation option in such cases.

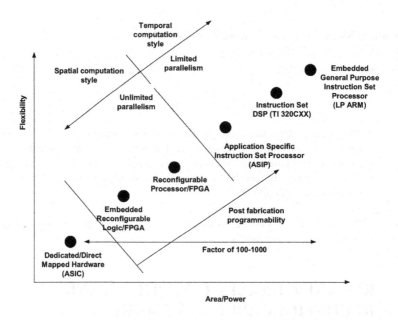

Figure 1-1. Positioning of reconfigurable hardware

There are also other reasons why to use reconfigurable resources in system-on-chip (SoC) design. The increasing non-recurring engineering (NRE) costs push designers to use same SoC in several applications and products for achieving low cost per chip. The presence of reconfigurable resources allows the fine tuning of the chip for different products or product variations. Also, the increasing complexity in the future designs adds the possibility of including design flows, which can require costly and slow redesign of the chip. Reconfigurable elements are often homogenous arrays, which can be pre-verified to minimize the possibility of having design errors. Also the post-manufacturing programmability allows correction or circumvention of problems later than with fixed hardware.

2. TYPES OF RECONFIGURATION

The next paragraphs describe different types of reconfiguration.

2.1 Logic reconfiguration

A typical logic block reconfigurable architecture contains a look-up table (LUT), an optional D flip-flop and additional combinational logic. The LUT allows any function to be implemented, providing generic logic. The flip-flop can be used for pipelining, registers, state holding functions for finite state machines, or any other situation where clocking is required. The combinatorial logic is usually the fast carry logic used to speed up fast carry-based computations such as addition, parity, wide AND operations and other functions. The logic blocks located at the periphery of the device can be of different architecture dedicated to I/O operations.

The logic blocks are grouped to matrices overlaid with a reconfigurable interconnection network of wires. Interconnection network reconfiguration is controlled by changing the connections between the logic blocks and the wires and by configuring the switch boxes, which connect different wires. The reconfiguration of both the logic blocks and the interconnection network is achieved by using SRAM memory bits to control the configuration of transistors. The functionality of the logic blocks, I/O blocks and the interconnection network is modified by downloading bit stream of reconfiguration data onto the hardware.

2.2 Instruction-set reconfiguration

The concept of instruction-set reconfiguration refers to the hybrid architectures consisting of microprocessor and reconfigurable logic. The key benefit is a combination of full software flexibility with high hardware efficiency. One promising approach is the reconfigurable instruction set processors (RISP), which have the capability to adapt their instruction sets to the application being executed through a reconfiguration in their hardware. The result is a reconfigurable and extensible processor architecture, which could be tailored closely to the designers' specific needs.

Through the adaptation, specialized hardware accelerates the execution of the applications. If shared resources are used in the adaptation, the functional density is also improved. By moving the application-specific data-paths into the processor, a remarkable improvement in performance compared to fixed instruction-set processors can be achieved. At the same time, designing at the level of instruction-set architecture significantly shortens the design cycle and reduces verification effort and risk. On the other hand, new methodologies, tools and processor foundations are required. Automated extension of processor function units and associated software environment - compilers, debuggers, instruction simulators etc., are also the key points to success.

Different systems with different characteristics have been designed. Usually two main design tasks are involved:

1. What is the type of interfaces between the microprocessor and the reconfigurable logic?
2. How to design the reconfigurable logic itself?

Each of them contains many trade-offs. The common classification of the reconfigurable processors could be made according to the coupling levels of reconfigurable logic. The concept of coupling levels applies also without a reference to reconfigurable processors. As shown in Figure 1-2, there are three types of coupling levels:

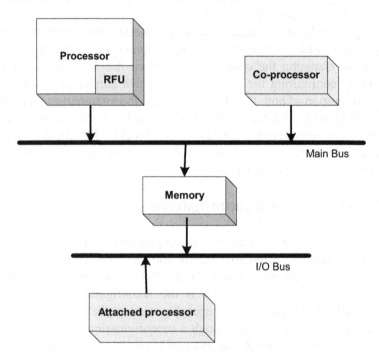

Figure 1-2. Basic coupling levels of reconfigurable logic

1. *Reconfigurable functional unit (RFU)* - the logic is placed inside the processor, the instruction decoder issues instructions to the reconfigurable unit as if it were one of the standard functional units of the processor. In this way, the communication cost is very small, so the speed could be easily increased. This is also the most promising

approach because it can be used to accelerate almost any application [1].

2. *Coprocessor* - the logic is next to the processor. Communication is done using a protocol.

3. *Attached processor* - the logic is placed on some kind of I/O bus. With the coprocessor and attached processor approaches, the speed improvement using the reconfigurable logic has to compensate for the overhead of transferring the data. This usually happens in applications where a huge amount of data has to be processed using a simple algorithm that fits in the reconfigurable logic.

2.3   Static and dynamic reconfiguration

There are two basic reconfiguration approaches: *static* and *dynamic*.

2.3.1 Static reconfiguration

Static reconfiguration (often referred as *compile time reconfiguration*) is the simplest and most common approach for implementing applications with reconfigurable logic. Static reconfiguration involves hardware changes at a relatively slow rate. It is a static implementation strategy where each application consists of one configuration. The main objective is to improve the performance.

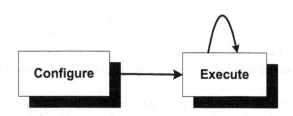

Figure 1-3. Principle of static reconfiguration

The distinctive feature of this configuration is that it consists of a single system-wide configuration. Prior to commencing an operation, the reconfigurable resources are loaded with their respective configurations. Once operation commences, the reconfigurable resources will remain in this configuration throughout the operation of the application. Thus hardware resources remain static for the life of the design (or application). This is depicted in Figure 1-3. Much higher performance than with pure software implementation (e.g. microprocessor approaches), cost advantage over

ASICs in certain cases and conventional CAD tool support are the main advantages of this technology.

2.3.2 Dynamic reconfiguration

Whereas static reconfiguration allocates logic for the duration of an application, dynamic reconfiguration (often referred to as *run time reconfiguration*) uses a dynamic allocation scheme that re-allocates hardware at run-time. This is an advanced technique that some people regard as a flexible realization of the time/space trade-off. It can increase system performance by using highly optimized circuits that are loaded and unloaded dynamically during the operation of the system as depicted in Figure 1-4. In this way system flexibility is maintained and functional density is increased [9].

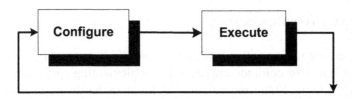

Figure 1-4. Principle of dynamic reconfiguration

Dynamic reconfiguration is based upon the concept of virtual hardware, which is similar to the idea of virtual memory. Here, the physical hardware is much smaller than the sum of the resources required by all of the configurations. Therefore, instead of reducing the number of configurations that are mapped, we instead swap them in and out of the actual hardware, as they are needed.

There are two main design problems for this approach: the first is to divide the algorithm into time-exclusive segments that do not need to (or cannot) run concurrently. This is referred to as temporal partitioning. Because no CAD tools support this step, this requires tedious and error-prone user involvement. The second problem is to co-ordinate the behaviour between different configurations, i.e. the management of transmission of intermediate results from one configuration to the next [8].

3. CLASSIFICATION OF RECONFIGURABLE ARCHITECTURES

In this section reconfigurable hardware architectures are classified with respect to several parameters. These parameters are described below:

- **Granularity of building blocks** This refers to the levels of manipulation of data. In this chapter we distinguish three types of granularity: *fine-grain* which corresponds to bit-level manipulation of data, *medium grain* manipulating data with varying number of bits and *coarse-grain* granularity which implies word level operations.
- **Reconfiguration scheme** Systems can be reconfigured statically or dynamically. Dynamically reconfigurable systems permit the partial reconfiguration of certain logic blocks while others are performing computations. Statically reconfigurable devices require execution interrupt.
- **Coupling** This refers to the degree of coupling with a host microprocessor. In a *closely coupled* system reconfigurable units are placed on the data path of the processor, acting as execution units. *Loosely coupled* systems act as a coprocessor. They are connected to a host computer system through channels or some special-purpose hardware.

3.1 Classification with respect to building blocks granularity

The granularity criterion reflects the smallest block of which a reconfigurable device is made.

In *fine-grained* architectures, the basic programmed building block usually consists of a combinatorial network and a few flip-flops. The logic block can be programmed into a simple logic function, such as a 2-bit adder. These blocks are connected through a reconfigurable interconnection network. More complex operations can be constructed by reconfiguring this network. Commercially available Field Programmable Gate Arrays (FPGAs) are based on fine grain architectures.

Although highly flexible, these systems exhibit a low efficiency when it comes to more specific tasks. For example, although an 8-bit adder can be implemented in a fine-grained circuit, it will be inefficient, compared to a reconfigurable array of 8-bit adders, when performing an addition-intensive task. An 8-bit adder will also occupy more space in the fine-grained implementation.

Reconfigurable systems which use logic blocks of larger granularity are categorized as *medium-grained* [6, 7, 10, 11, 17]. For example, Garp [6] is designed to perform a number of different operations on up to four 2-bit inputs. Another medium-grained structure was designed specifically to implement multipliers of a configurable bit-width [7]. The logic block used in the multiplier FPGA is capable of implementing a 4x4 multiplication. The CHESS architecture [11] also operates on 4-bit values, with each of its cells acting as a 4-bit ALU. The major advantage of medium-grained systems when compared to the fine-grained architecture is, that they better utilize the chip area, since they are optimized for the specific operations. However, a drawback of this approach is represented in a high overhead when synthesizing operations which are incompatible with the simplest logic block architecture.

Coarse-grained architectures are primarily intended for the implementation of tasks dominated by word-width operations. Because the logic blocks used are optimized for large computations, they will perform these operations much more quickly (and consume less chip area) than a set of smaller cells connected to form the same type of structure. However, because their composition is static, they are unable to leverage optimizations in the size of operands. On the other hand, these coarse-grained architectures can be much more efficient than finer-grained architectures for implementing functions closer to their basic word size. An example of coarse-grained system is the RaPiD architecture [4].

A *very coarse granularity* is the case when the simplest logic block is based on an entire microprocessor with or without special accelerators. Examples of such architectures are the REMARC [12] and RAW [13] architectures.

3.2   Classification with respect to reconfiguration scheme

3.2.1   Statically reconfigurable architectures

Traditional reconfigurable architectures are *statically reconfigurable*, which means that the reconfigurable resources are configured at the start of execution and remain unchanged for the duration of the application. In order to reconfigure a statically reconfigurable architecture, the system has to be halted while the reconfiguration is in progress and then restarted with the new configuration.

Traditional FPGA architectures have primarily been serially programmed single-context devices, allowing only one configuration to be loaded at a time. This type of FPGAs is programmed using a serial stream of configuration information, requiring a full reconfiguration if any change is required.

3.2.2 Dynamically reconfigurable architectures

Dynamically reconfigurable (run-time reconfigurable) architectures allow reconfiguration and execution to proceed at the same time. The different reconfigurable styles of dynamic reconfiguration are depicted in Figure 1-5 and discussed in the following paragraphs.

Single context dynamically reconfigurable architectures

Although single context architectures can typically be reconfigured only statically, a run-time reconfiguration onto single context FPGA can also be implemented. Typically, the configurations are grouped into contexts, and each context is swapped as needed. Attention has to be paid on proper partitioning of the configurations between the contexts in order to minimize the reconfiguration delay.

Multi-context dynamically reconfigurable architectures

A multi-context architecture includes multiple memory bits for each programming bit location. These memory bits can be thought of as multiple planes of configuration information [3, 15]. Only one plane of configuration information can be active at a given moment, but the architecture can

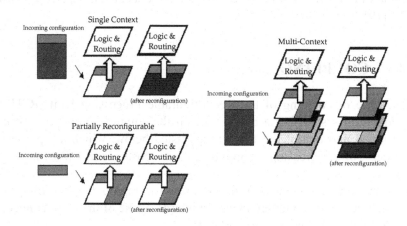

Figure 1-5. The different basic models of dynamically reconfigurable computing

quickly switch between different planes, or contexts, of already-programmed configurations. In this manner, the multi-context architecture can be considered a multiplexed set of single-context architectures, which requires that a context be fully reprogrammed to perform any modification to the configuration data. However, this requires a great deal more area than the other structures, given that there must be as many storage units per programming location as there are contexts. This also means that multi-context schemes are mainly used in coarse-grain architectures.

Partially Reconfigurable Architectures

In some cases, configurations do not occupy the full reconfigurable hardware, or only a part of a configuration requires modification. In both of these situations a partial reconfiguration of the reconfigurable resources is desired, rather than the full reconfiguration supported by the serial architectures mentioned above.

In partially reconfigurable architectures, the underlying programming layer operates like a RAM device. Using addresses to specify the target location of the configuration data allows for selective reconfiguration of the reconfigurable resources. Frequently, the undisturbed portions of the reconfigurable resources may continue execution, allowing the overlap of computation with reconfiguration. When configurations do not require the entire area available within the array, a number of different configurations may be loaded into otherwise unused areas of the hardware. Partially run-time reconfigurable architectures can allow for complete reconfiguration flexibility such as the Xilinx 6200 [18], or may require a full column of configuration information to be reconfigured at once, as in the Xilinx Virtex FPGA [19].

4.   COUPLING

The type of coupling of the Reconfigurable Processing Unit (RPU) to the computing system has a big impact on the communication cost. It can be classified into one of the four groups listed below, which are presented in order of decreasing communication costs and illustrated in Figure 1-6:

- RPUs coupled to the I/O bus of the host (Figure 1-6.a). This group includes many commercial circuit boards. Some of them are connected to the PCI bus of a PC or workstation.
- RPUs coupled to the local bus of the host (Figure 1-6.b).

- RPUs coupled like co-processors (Figure 1-6.c) such as the REMARC - Reconfigurable Multimedia Array Coprocessor [12].
- RPUs acting like an extended data-path of the processor (Figure 1-6.d) such as the OneChip [16], the PRISC - Programmable Reduced Instruction Set Computer [14], and the Chimaera [5].

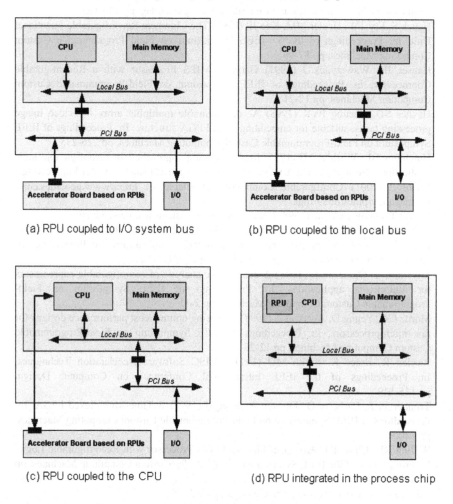

(a) RPU coupled to I/O system bus

(b) RPU coupled to the local bus

(c) RPU coupled to the CPU

(d) RPU integrated in the process chip

Figure 1-6. Coupling of the RPU to the host computer

REFERENCES

1. Barat F, Lauwereins R (2000) Reconfigurable Instruction Set Processors: A Survey. In: Proceedings of IEEE international Workshop on Rapid System Prototyping, pp 168-173

2. Brodersen B (2002) Wireless Systems-on-a-Chip Design. In: Proceedings of 3rd International Symposium on Quality of Electronic Design, pp 221-222

3. DeHon A (1996) DPGA Utilization and Application. In: Proceedings of ACM/SIGDA International Symposium on FPGAs, pp 115-121

4. Ebeling C, Cronquist DC, Franklin P (1996) RaPiD Reconfigurable Pipelined Datapath. In: Lecture Notes in Computer Science 1142 – Field Programmable Logic: Smart Applications, New Paradigms and Compilers, Springer Verlag, pp 126-135

5. Hauck S, Fry TW, Hosler MM, Kao JP (1997) The Chimaera Reconfigurable Functional Unit. In: Proceedings of the 5th IEEE Symposium on Field Programmable Custom Computing Machines, pp 87-96

6. Hauser JR, Wawrzynek J (1997) Garp: A MIPS Processor with a Reconfigurable Coprocessor. In: Proceedings of IEEE Symposium on Field-Programmable Custom Computing Machines, pp 12-21

7. Haynes SD, Cheung PYK (1998) A reconfigurable multiplier array for video image processing tasks, suitable for embedding in an FPGA structure. In: Proceedings of IEEE Symposium on Field-Programmable Custom Computing Machines, pp 226-235

8. Hutchings BL, Wirthlin MJ (1995) Implementation approaches for reconfigurable logic applications. Brigham Young University, Dept. of Electrical and Computer Engineering

9. Khatib J (2001) Configurable Computing. Available at: http://www.geocities.com/ siliconvalley/pines/6639/fpga

10. Lucent Technologies Inc (1998) FPGA Data Book, Allentown, Pennsylvania

11. Marshall A, Stansfield T, Kostarnov I, Vuillemin J, Hutchings B (1999) A Reconfigurable Arithmetic Array for Multimedia Applications. In: Proceedings of ACM/SIGDA International Symposium on FPGAs, pp 135-143

12. Miyamori T, Olukotun K (1998) A quantitative analysis of reconfigurable coprocessors for multimedia applications. In: Proceedings of IEEE Symposium on Field-Programmable Custom Computing Machines, pp 2-11

13. Moritz CA, Yeung D, Agarwal A (1998) Exploring optimal cost performance designs for raw microprocessors. In: Proceedings of IEEE Symposium on Field-Programmable Custom Computing Machines, pp 12-27

14. Razdan R, Brace K, Smith MD (1994) PRISC Software Acceleration Techniques. In: Proceedings of the IEEE International Conference on Computer Design, pp 145-149

15. Trimberger S, Carberry D, Johnson A, Wong J (1997) A Time-Multiplexed FPGA. In: Proceedings of IEEE Symposium on Field-Programmable Custom Computing Machines, pp 22-29

16. Witting RD, Chow P (1996) OneChip: An FPGA Processor with Reconfigurable Logic. In: Proceedings of the IEEE Symposium on FPGAs for Custom Computing Machines, pp 126-135

17. Xilinx Inc. (1994) The Programmable Logic Data Book

18. Xilinx Inc. (1996) XC6200: Advanced product specification v1.0. In: The Programmable Logic Data Book

19. Xilinx Inc. (1999) VirtexTM: Configuration Architecture Advanced Users' Guide

Chapter 2

RECONFIGURABLE HARDWARE EXPLOITATION IN WIRELESS MULTIMEDIA COMMUNICATIONS

Konstantinos Masselos[1,2] and Nikolaos S. Voros[1]

[1] *INTRACOM S.A., Hellenic Telecommunications and Electronics Industry, Greece*
[2] *Currently with Imperial College of Science Technology and Medicine, United Kingdom*

Abstract: This chapter presents cases where reconfigurable hardware can be exploited for the efficient realization of wireless multimedia communication systems. The various scenarios described are referring to (a) the DLC/MAC layer and the baseband part of the physical layer of HIPERLAN/2 and IEEE 802.11a WLAN protocols, and (b) the application layer of a sophisticated personal device. The goal of this chapter is to provide an insight on the advantages reconfigurable hardware may bring in real life applications.

Key words: Reconfiguration, WLAN, application layer, wireless multimedia communications

1. RECONFIGURABLE HARDWARE BENEFITS FROM A SYSTEM'S PERSPECTIVE

The presence of reconfigurable hardware resources in a system can be exploited in two major directions:

- To create space for post-fabrication functional modifications e.g. to upgrade system functionality or for software like bug fixing. Software realizations allow post-fabrication functional modifications, however for complex tasks software realizations might be inefficient. This feature may allow important time-to-market improvement.
- To allow sharing of hardware resources among tasks that are not active simultaneously thus reducing the total area cost of the system. Such

27

N.S. Voros and K. Masselos (eds.), System Level Design of Reconfigurable Systems-on-Chips, 27-42.
© 2005 *Springer. Printed in the Netherlands.*

tasks may belong to different modes of operation of a given system, to different applications or standards realized on the same platform or even to time non-overlapping tasks of a single system.

Given an application, tasks that are suitable for realization on reconfigurable hardware are those that may share hardware resources with other tasks over time or are likely to be modified/upgraded in the future and also have high computational complexity (that prevents efficient realization on instruction set processors).

In the rest of this chapter, reconfiguration scenarios are discussed from the wireless communications and multimedia domains. Real life complex systems are used for this analysis namely the HIPERLAN/2 and IEEE 802.11a WLAN systems (covering MAC and physical layers functionality) and the MPEG system (covering the application layer).

2.   RECONFIGURATION SCENARIOS FOR HIPERLAN/2 AND IEEE 802.11a WLAN SYSTEMS

In this section reconfiguration scenarios for the HIPERLAN/2 and IEEE 802.11a WLAN systems are discussed. The two systems targeted functionalities cover the DLC/MAC layer and the baseband part of the physical layer.

2.1   HIPERLAN/2 and IEEE 802.11a systems

HIPERLAN/2 [1] is a connection-oriented time-division multiple access (TDMA) system. Physical layer is based on coded OFDM modulation scheme [2]. The physical layer is multi-rate type allowing control of link capability between access point and mobile terminal according interference situations and distance.

The flow graph of the HIPERLAN/2 transmitter is shown in Figure 2-1. The blocks in the inputs and outputs of the different tasks give the input and output rates of the tasks respectively. The input rate of a given task corresponds to the minimum amount of data required for the task to produce a given output (output rate).

The computational complexity and the type of processing of the transmitter tasks are analytically presented in Table 2-1. The analysis of computational complexity is done by estimating the number of required basic operations per output data item in each function. The basic operations include arithmetic, logic and memory read/write operations. It is assumed, that a processing of transmitted or received data should be possible at a sustained nominal data rate of each physical layer mode. The input and output operations included in this complexity analysis correspond to data coming from previous tasks and being passed to following tasks (in a real implementation these operations are likely corresponding to accesses to data storage locations).

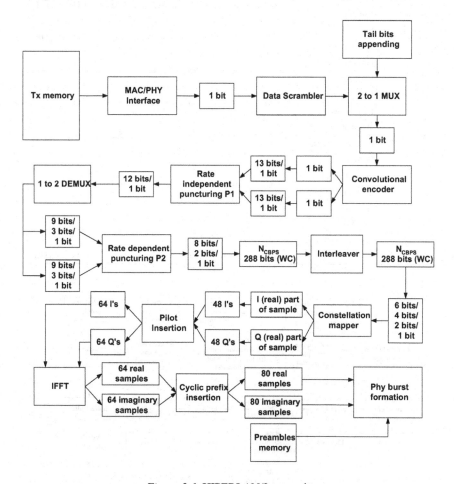

Figure 2-1. HIPERLAN/2 transmitter

From the computational complexity analysis it can be seen that there are some algorithms that generate a constant computational complexity in all physical layer modes. The most important is IFFT that is dominating the overall transmit side complexity in the low bit rate modes. The complexities of channel coding functions are naturally related to the used bit rate.

Table 2-1. Computational complexity of transmitter tasks in different physical layer modes

Task	Type of processing	Computational complexity (MOPS) / PHY mode (Mb/s)						
		6	9	12	18	27	36	54
Scrambling	bit level - shift register, XOR	108	162	216	324	486	648	972
Convolutional encoding	bit level - shift register, XOR	174	261	348	522	783	1044	1566
Puncturing (Rate dependent)	bit level – logic operations	0.31	0.31	0.31	0.31	0.31	0.31	0.31
Puncturing (Rate dependent)	bit level – logic operations	0	33	0	66	105	132	198
Interleaving	Group of bits – LUT accesses	48	48	96	96	192	192	288
Constellation mapping	Group of bits – LUT accesses	30	45	36	54	54	72	90
Pilot insertion	Word level - memory accesses	56	56	56	56	56	56	56
IFFT	Word level – multiplications, additions, memory accesses	922	922	922	922	922	922	922
Cyclic prefix insertion	Word level - memory accesses	72	72	72	72	72	72	72
Sum		1410	1599	1746	2112	2670	3138	4164

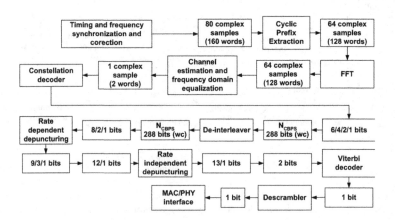

Figure 2-2. HIPERLAN/2 receiver

The flow graph of a reference HIPERLAN/2 receiver is presented in Figure 2-2. The receiver chain of the HIPERLAN/2 is left open by the standard so there is more freedom for algorithm selection for certain blocks such as the timing and frequency synchronization and the channel estimation (different chains of tasks can be adopted for these two generic blocks). The computational complexity and the type of processing of the receiver tasks are analytically presented in Table 2-2.

Table 2-2. Computational complexity of receiver tasks in different physical layer modes

Task	Type of processing	Computational complexity (MOPS) / PHY mode (Mb/s)						
		6	*9*	*12*	*18*	*27*	*36*	*54*
Cyclic prefix extraction	*Word level– memory accesses*	96	96	96	96	96	96	96
Frequency error correction	*Word level – multiplications, additions, memory accesses*	208	208	208	208	208	208	208
FFT	*Word level – multiplications, additions, memory accesses*	922	922	922	922	922	922	922
Frequency domain equalization	*Word level – multiplications, additions, memory accesses*	132	132	132	132	132	132	132
Constellation demapping	*Group of bits – LUT accesses*	48	48	240	240	288	288	336
Deinterleaving	*Group of bits – LUT accesses*	48	48	96	96	192	192	288
Depuncturing (Rate dependent)	*bit level – logic operations*	0	50	0	99	118	198	297
Depuncturing (Rate independent)	*bit level – logic operations*	0.16	0.20	0.16	0.20	0.28	0.20	0.20
Viterbi decoding	*Bit level I/O – word level additions, comparisons*	1170	1755	2340	3510	5265	7020	10530
Descrambling	*bit level – shift register, XOR*	108	162	216	324	486	648	972
Sum		2732	3421	4250	5627	7707	9704	13781

As it can be deduced, the Viterbi decoding dominates the overall complexity figures in all physical layer modes. It can be also seen that the receiver side processing is up to three times more complex than transmit side processing.

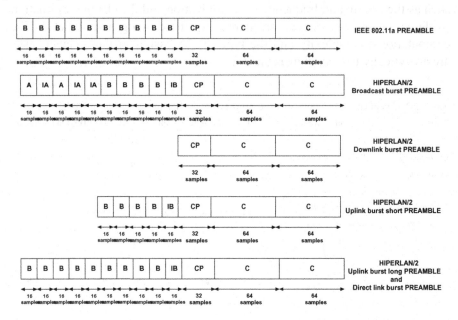

Figure 2-3. IEEE 802.11a and HIPERLAN/2 preambles

The baseband part of the IEEE 802.11a system [3] is almost similar to that of HIPERLAN/2 system. Only some minor differences exist. IEEE 802.11a uses only one preamble sequence (shown in Figure 2-3) of 320 samples. HIPERLAN/2 uses 4 different types of preamble sequences for the different types of PDUs with sizes ranging from 160 samples to 320 samples. The contents of the first half of the PREAMBLE sequences of HIPERLAN/2 are always different to that of IEEE 802.11a. From an implementation point of view this may affect the synchronization block of the receiver.

Different sequences are used by the two systems for the initialization of the (de)scrambler. In IEEE 802.11a the initialization is performed using the first 7 bits of the service field which are always set to zero. In HIPERLAN/2 the initial state of the scrambler is set to pseudo random non-zero 7-bit state determined by the frame counter field in the BCH (first four bits of BCH) at the beginning of the corresponding MAC frame. The initial state is derived by appending the first four bits of BCH to the fixed binary number $(111)_2$. This difference is small from an implementation point of view.

In the encoder side, IEEE 802.11a supports 1/2, 3/4 and 2/3 code rates while HIPERLAN/2 supports 1/2, 3/4 and 9/16 code rates. Two code rates are in common while each system supports a third different extra one. HIPERLAN/2 applies two puncturing stages (a rate independent one followed by a rate dependent one) while IEEE 802.11a applies a single puncturing stage. The puncturing patterns applied by the two systems to achieve the different code rates are presented in Figure 2-4 (no puncturing pattern is required for 1/2 code rate). The difference from an implementation point of view is small.

The combinations of modulation, coding rate and achieved nominal bit rate (physical modes of operation) supported by IEEE 802.11a and HIPERLAN/2 are presented in Table 2-3. Six modes of operation are common, IEEE 802.11a supports two extra modes while HIPERLAN/2 supports one extra mode. From an implementation point of view the number of modes of operation supported affects the modem controller from which the modem control words are issued.

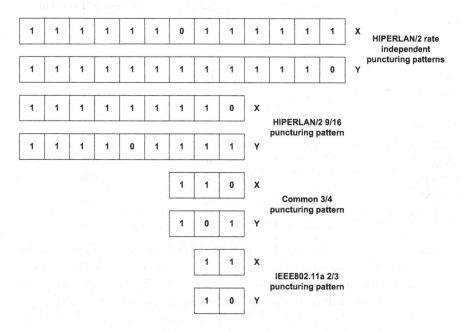

Figure 2-4. Puncturing patterns used by IEEE 802.11a and HIPERLAN/2

The MAC frame duration of the HIPERLAN/2 is fixed to 2 ms. The HIPERLAN/2 MAC frame structure described in Figure 2-5 comprises time slots for broadcast control (BCH), frame control (FCH), access feedback control (ACH) and data transmission in downlink (DL), uplink (UL) and directlink (DiL) phases, which are allocated dynamically depending on the need for transmission resources. A mobile terminal (MT) first has to request capacity from the access point (AP) in order to send data. This can be done in the random access channel (RCH), where contention for the same time slot is allowed. Downlink, uplink and directlink phases consist of two types of PDUs. The long PDUs have a size of 54 bytes and contain control or user data. The payload is 49.5 bytes and the remaining 4.5 bytes are used for the PDU Type (2 bits), a sequence number (10 bits, SN) and cyclic redundancy check (CRC-24). Long PDUs are referred to as the long transport channel (LCH). Short PDUs contain only control data and have a size of 9 bytes. They may contain resource requests, ARQ messages etc and they are referred to as the short transport channel (SCH). A physical burst is composed of the PDU train payload and a preamble and is the unit to be transmitted via the physical layer.

Table 2-3. Physical modes of operation of IEEE 802.11a and HIPERLAN/2

Modulation	Coding Rate R	Nominal bit rate (Mbit/s)	Coded bits per OFDM symbol
BPSK	1/2	6	48
BPSK	3/4	9	48
QPSK	1/2	12	96
QPSK	3/4	18	96
16 QAM (HL/2 only)	9/16	27	192
16 QAM (IEEE 802.11a only)	1/2	24	192
16 QAM	3/4	36	192
64 QAM	3/4	54	288
64 QAM (IEEE 802.11a only)	2/3	48	288

The structure of the IEEE 802.11a PPDU frame is described in Figure 2-6. The header contains information about the length of the exchanged data and the transmission rate. The RATE field conveys information about the type of the modulation and the coding rate used in the rest of the packet. The LENGTH field takes a value between 1 and 4095 and specifies the number of bytes to be exchanged (PSDU). The six tail bits are used to reset the convolutional encoder and to terminate the code trellis in the decoder. The first 7 bits of the service field are set to zero and are used to initialise the (de)scrambler. The remaining 9 bits are reserved for future use.

The pad bits are used to ensure that the number of bits in the PPDU frame maps to an integer number of OFDM symbols. A cyclic redundancy check (CRC-32) is included in the IEEE 802.11a PSDU.

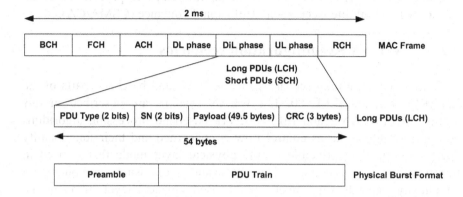

Figure 2-5. HIPERLAN/2 MAC frame, Long PDU and Physical Burst format

An important issue is that the transmission duration (TXTIME) for a PPDU frame in IEEE 802.11a is not fixed but a function of LENGTH field as shown in the following equation:

$$TXTIME = T_{PREAMBLE} + T_{SIGNAL} + T_{SYM} \times Ceiling((16 + 8 \times LENGTH + 6) / N_{DBPS}) \qquad (1)$$

where N_{DBPS} is the number of data bits per symbol and can be derived from the DATARATE parameter. From an implementation point of view this fact imposes a strict timing requirement to the MAC/PHY interface for the decoding of the SIGNAL symbol in order to determine the number of OFDM symbols to be exchanged.

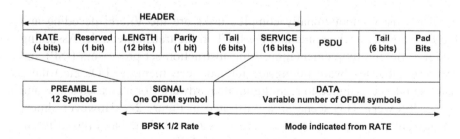

Figure 2-6. IEEE 802.11a PPDU frame format

The major differences between IEEE 802.11a and HIPERLAN/2 systems occur in the MAC sublayer. In HIPERLAN/2 the medium access is based on a TDD/TDMA approach. The control is centralized to an AP, which informs the MTs at which point in time in the MAC frame they are allowed to transmit their data. IEEE 802.11a uses a distributed MAC protocol based on Carrier Sense Multiple Access with Collision Avoidance (CSMA/CA).

2.2   WLAN Reconfiguration scenarios

Some reconfiguration scenarios for the MAC and baseband parts of the HIPERLAN/2 and IEEE 802.11a WLAN systems are described in this section. HIPERLAN/2 and IEEE 802.11a baseband processing algorithms are quite simple as far as control flow is concerned and their functionality does not depend in principle on the physical layer mode that is used in transmission or reception. The baseband processing computational complexity depends very much on the used physical layer mode in the transmission or reception.

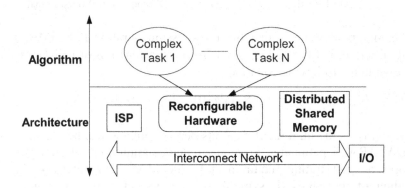

Figure 2-7. Realization on a highly flexible platform

The most computationally complex tasks are the Viterbi decoding and the FFT on the receiver side and the IFFT in the transmitter side. Assuming a highly flexible implementation using instruction set processors (ISP) and reconfigurable hardware (alongside interconnect, memory, I/Os etc.) these tasks should be assigned to reconfigurable hardware (for increased speed and reduced power). This scenario is illustrated in Figure 2-7. However almost no flexibility is required for these tasks on a stand-alone basis (no different candidate implementation choices exist). If ASIC blocks were included in the target implementation platform these tasks should be preferably moved to them.

Reconfigurable hardware resources can be shared among baseband processing tasks that are not active simultaneously. This may lead to silicon area optimization (taking into consideration reconfiguration related overheads). This scenario is described in Figure 2-8. For example under a half duplexing scenario the transmitter and the receiver will not be active simultaneously. In this case, tasks of the transmitter and the receiver may share the same reconfigurable hardware resources.

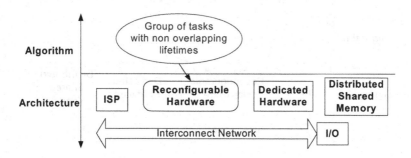

Figure 2-8. Reconfigurable hardware sharing among tasks with non-overlapping lifetimes

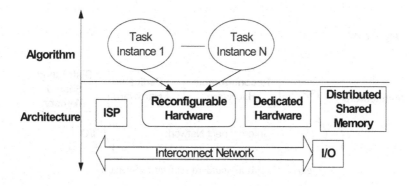

Figure 2-9. Realization of different algorithmic instances of the same task on reconfigurable hardware

Certain tasks in the receiver chain of the baseband processing allow different algorithmic implementations with different trade-offs between algorithmic performance and computational complexity (e.g. channel estimation). Lower algorithmic performance requirements (e.g. SNR, BER) may allow the use of less sophisticated and computational complex algorithmic instances leading to improved implementation efficiency (speed, power). Furthermore realization of different algorithmic instances for the same task in a given system may be beneficial e.g. allowing adaptation to different operating conditions. Such tasks are good candidates for implementation on reconfigurable hardware (with their different instances sharing the same reconfigurable hardware resources) if their complexity is high (preventing efficient realization on instruction set processors). This scenario is described in Figure 2-9.

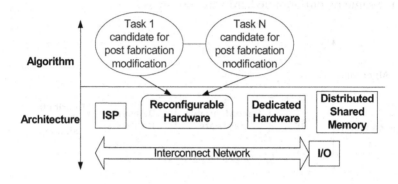

Figure 2-10. Post shipment modification scenario

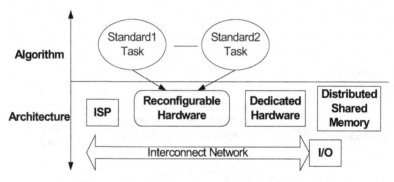

Figure 2-11. Multi-standard realization scenario

Another opportunity for reconfigurable hardware exploitation is towards post-shipment modification/enhancement of the system's functionality (e.g. with more sophisticated realizations of certain tasks). Baseband processing tasks that are candidates for being upgraded are those that are left open by the standard. This scenario is described in Figure 2-10.

More opportunities for reconfiguration and reconfigurable hardware sharing exist in the case of realization of multiple standards on the same reconfigurable implementation platform. This scenario is described in Figure 2-11. Let assume a HIPERLAN/2 – IEEE 802.11a dual standard realization with the two systems not being active simultaneously. Given that the major differences between the two standards are in the MAC layers reconfigurable hardware can be used for the realization of the most complex and performance demanding parts of the MAC layers (and the MAC to baseband interfaces) of the two systems.

3. RECONFIGURATION SCENARIOS AT THE APPLICATION LAYER

As portable devices become more powerful, it also becomes possible to run more computationally intensive services on these appliances. Due to the increasing flexibility requirements that are imposed by these applications, the devices need to be highly adaptable to the running applications. At the other hand, efficient realizations of these applications are required, especially in the resources they use during deployment, where power consumption must be traded against perceived quality of the application. To be able to realize a variety of applications or services, the implementation platform needs to be highly adaptable.

Assume a wireless communication terminal as is shown in Figure 2-12, which consists out of instruction set processors (ISP) and reconfigurable hardware that are connected to a common interconnect network and to memory. This device is powerful enough to support various applications, including video. Because of the high computational demand of such a video application, it will be run on the reconfigurable hardware (see Figure 2-12) as that part can be configured for optimal performance for a given application.

When the user decides to view the video in a small window and to start up a 3D game, the situation changes. Then the video application can be run with much less resources, while the game becomes the most computationally intensive application. This means that this 3D game will need to be run on the reconfigurable hardware. To enable that, the video application is moved to run further in software on an instruction set processor (ISP). The hardware is then reconfigured for the 3D game and that application is started (see Figure 2-13).

By moving the video application to software and running it in a smaller window also implies that a lower data rate can be used on the wireless terminal interconnect. This means that the wireless appliance should send back to the server that a lower resolution (and thus a lower bit-rate) is allowed for the video application. The application quality as perceived by the user is still satisfying.

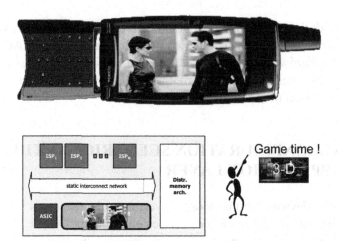

Figure 2-12. A video application is running on the reconfigurable hardware

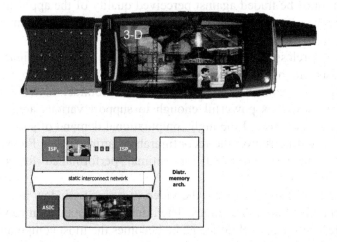

Figure 2-13. A 3D application is running on the reconfigurable hardware, while the video application continues in a reduced window and on a software processor

From the application scenario above, it is clear that it must be possible to run many different applications on the reconfigurable hardware. This means that general reconfigurable hardware is needed, in contrast to incorporating dedicated hardware blocks, like FFT processor, FIR filter etc. Also we notice that applications are very different in nature, as already described in the case of video streaming and interactive 3D applications. A selection of the reconfiguration characteristics is also based on general characteristics of the multi-media applications and on the usage scenario above.

Requirements on reconfiguration time are modest: because reconfiguration is user-initiated, fast reconfiguration times (< 1 msec) are not needed. When e.g. switching a video application from hardware to software, it is not important that a numbers of frames are not decoded. As soon as the application is running in software, it decodes the next incoming frame.

Requirements on the reconfiguration granularity are complicated by the unknown nature of the application, the granularity should be fine enough so that for each application an optimal implementation in reconfigurable hardware is possible. However due to power requirements, word level coarse grain reconfiguration is more appropriate than bit-level reconfiguration. This is especially the case when the word-lengths are matched to the application at hand.

Table 2-4. Operational power requirements for MPEG2 video decoding

MPEG-2 MP@ML Decoder			
Function	*MOPS*	*Input*	*Output*
Bitstream parsing and VLD	12	4	40
Dequantization and IDCT	105	40	70
Motion Compensation	273	70	70
YUV to RGB color conversion	299	70	35
Total	689	184	215

Table 2-5. Operational power requirements for a 3D application

Quality	CPU time	#triangles	#pixels	Architecture
31 dB	40 ms	5000	5 %	SW
31 dB	2 ms	5000	5 %	HW
25 dB	70 ms	5000	19%	SW
30 dB	80 ms	8000	19%	SW
43 dB	118 ms	17500	19 %	SW
43 dB	21 ms	17500	19 %	HW

To summarize the requirements on applications, it is not only emphasized that different applications must be able to run on the wireless LAN platform, but also that they can have huge computational demands for which dedicated or reconfigurable hardware is needed. To have an indication of the required operational power, we refer to literature [4, 5] the results of which are summarized in Table 2-4 for MPEG2 and in Table 2-5 for a 3D application. In the latter application the CPU time, and thus the frame rate, is closely related to the required quality (application QoS) but also depends on the architecture, be it a hardware or a software realization.

REFERENCES

1. ETSI (2000), Broadband Radio Access Networks (BRAN); HIPERLAN type 2; Physical (PHY) layer, v1.2.1
2. Van Nee R, Prasad R (1999) OFDM for Mobile Multimedia Communications. Boston: Artech House
3. IEEE Std 802.11a/D7.0 (1999) Part 1: Wireless LAN Medium Access Control (MAC) and Physical Layer (PHY) specifications: High Speed Physical Layer in the 5 GHz Band
4. Zhou CG, Kabir I, Kohn L, Jabbi A, Rice D, Hu XP (1995) MPEG video decoding with the UltraSPARC visual instruction set. In: Proceedings of the 40th IEEE Computer Society International Conference, pp. 470 – 477
5. Lafruit G, Nachtergaele L, Denolf K, Bormans J (2000) 3D Computational Graceful Degradation. In: Proceedings of ISCAS – Workshop and Exhibition on MPEG-4, vol. 3, pp. 547-550

Chapter 3

RECONFIGURABLE HARDWARE TECHNOLOGIES

Konstantinos Masselos[1,2] and Nikolaos S. Voros[1]

[1] *INTRACOM S.A., Hellenic Telecommunications and Electronics Industry, Greece*
[2] *Currently with Imperial College of Science Technology and Medicine, United Kingdom*

Abstract: A large number of reconfigurable hardware technologies have been proposed both in academia and commercially (some of them in their first market steps). They can be roughly classified in three major categories: a) Field Programmable Gate Arrays (FPGAs), b) integrated circuit devices with embedded reconfigurable resources and c) embedded reconfigurable cores for Systems-on-Chip (SoCs). In this chapter representative commercial technologies are discussed and their main features are presented[1].

Key words: Field Programmable Gate Arrays (FPGAs), embedded reconfigurable cores, fine grain reconfigurable architecture, coarse grain reconfigurable architecture

1. FIELD PROGRAMMABLE GATE ARRAYS (FPGAS)

Field programmable gate arrays currently represent the most popular and mature segment of reconfigurable hardware technologies. Technology advances keep increasing the gates counts and memory densities of FPGAs while they also allow the integration of functions ranging from hardwired multipliers through high speed transceivers and all the way up to (hard or soft) CPU cores with associated peripherals. These advances make possible the realization of complete systems on a single FPGA chip improving end-system size, power consumption, performance, reliability and cost. Equally

---

[1] The information included in this chapter is up-to-date until November 2004.

N.S. Voros and K. Masselos (eds.), System Level Design of Reconfigurable Systems-on-Chips, 43-83.

important FPGAs can be reconfigured in seconds either statically or dynamically/partially. Reconfiguration can take place in the workstation, in the assembly line or even at the end user premises. These capabilities provide flexibility:

- to react to last minute design changes
- to prototype ideas before implementation
- to meet time-to-market deadlines
- to correct errors and upgrade functions once the end system is in users' hands
- or even to implement reconfigurable computing i.e. using a fixed number of logic gates to time-division-multiplex multiple functions.

Because of all these advantages, FPGAs have been making significant inroads into ASIC territory. It is a matter of the per-gate cost decreases and the gates per device increases to decide whether FPGAs can replace ASICs.

Mapping of applications on FPGAs has been based on VHDL and Verilog languages for input descriptions. C based approaches are also currently under consideration. The integration of CPUs on FPGAs introduced design flows and tools supporting hardware/software codesign and software development.

There are a number of companies building FPGAs including Actel, Altera, Atmel, Lattice Semiconductor, Quicklogic and Xilinx; Xilinx and Altera currently being the market leaders. In order to differentiate, FPGA vendors have introduced devices to address different intersections of performance, power, integration and cost targets. Some representative FPGA devices are briefly discussed in the following subsections.

1.1   ALTERA Stratix II

Altera claims that Stratix II devices [11] are industry's fastest and highest density FPGAs. Stratix II devices extend the possibilities of FPGA design, allowing designers to meet the high-performance requirements of today's advanced systems and avoid developing with costly ASICs.

1.1.1   Architecture

The Stratix II architecture has been designed to primarily optimize performance but also logic density in a given silicon area. Its logic structure is constructed with Altera's new adaptive logic modules (ALMs). The Stratix II architecture reduces significantly the logic resources required to implement any given function and the number of logic levels in a given critical path. The architecture accomplishes this by permitting inputs to be shared by adjacent look-up tables in the same ALM. Multiple, independent functions can also be packed into a single ALM, further reducing interconnect delays and logic resource requirements. The structure of a Stratix II ALM is shown in Figure 3-1.

Stratix II FPGAs utilize the TriMatrix memory structure. TriMatrix memory includes the 512-bit M512 blocks, the 4-Kbit M4K blocks, and the 512-Kbit M-RAM blocks, each of which can be configured to support a wide range of features. Each embedded RAM block in the TriMatrix memory structure targets a different class of applications: the M512 blocks can be used for small functions such as first-in first-out (FIFO) applications, the M4K blocks can be used to store incoming data from multi-channel I/O protocols, and the M-RAM blocks can be used for storage-intensive applications such as Internet protocol packet buffering or program/ data memory for an on-chip Nios embedded processor. All memory blocks include extra parity bits for error control, embedded shift register functionality, mixed-width mode, and mixed-clock mode support. Additionally, the M4K and M-RAM blocks support true dual-port mode and byte masking for advanced write operations.

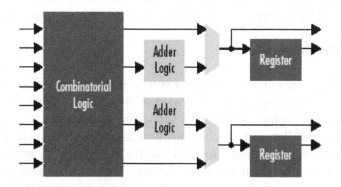

Figure 3-1. Stratix II adaptive logic module structure

Stratix II DSP blocks are optimized to implement processing intensive functions such as filtering, transforms, and modulation. Capable of running at 370 MHz, Stratix II DSP blocks provide maximum DSP throughput (up to 284 GMACs) that is orders of magnitude higher than leading-edge digital signal processors available today. Each DSP block can support a variety of multiplier bit sizes (9x9, 18x18, 36x36) and operation modes (multiplication, complex multiplication, multiply-accumulate and multiplyadd) and can generate DSP throughput of 3.0 GMACS per DSP block. In addition, rounding and saturation support has been added to the DSP block.

Stratix II FPGAs support many high-speed I/O standards and high-speed interfaces such as 10 Gigabit Ethernet (XSBI), SFI-4, SPI 4.2, HyperTransport™, RapidIO™, and UTOPIA Level 4 interfaces at up to 1 Gbps. These allow interfacing with anything from backplanes, host processors, buses and memory devices to 3D graphics controllers.

Stratix II devices support internal clock frequency rates of up to 500 MHz and typical design performance at over 250 MHz. Logic densities of Stratix II devices range from 15,600 to 179,400 equivalent logic elements. Total memory densities can be up to 9 Mbits of RAM, which can be clocked at a 370 MHz maximum clock speed. Stratix II FPGAs may include up to 12 PLLs and up to 48 system clocks per device.

1.1.2   Granularity

Stratix II architecture is a fine grain architecture with embedded hardwired word level modules.

1.1.3   Technology

Stratix II FPGAs are manufactured on 300-mm wafers using TSMC's 90-nm, 1.2-V, all-layer copper SRAM, low-k dielectric process technology.

1.1.4   Reconfiguration

Stratix II devices are configured at system power-up with data stored in an Altera configuration device or provided by an external controller. The Stratix II device's optimized interface allows microprocessors to configure it serially or in parallel, and synchronously or asynchronously. The interface also enables microprocessors to treat Stratix II devices as memory and configure them by writing to a virtual memory location, making reconfiguration easy. Remote system upgrades can be transmitted through any communications network to Stratix II devices.

1.1.5   Other issues

- Nios embedded processors allow designers to integrate embedded processors on Stratix II devices for complete system-on-a-programmable-chip (SOPC) designs. The Nios soft embedded processor has been optimized for the advanced architectural features of the Stratix II device family.

- Stratix II family enables design security through non-volatile, 128-bit AES design encryption technology for preventing intellectual property theft.
- A seamless, cost-reduction migration path to low-cost HardCopy structured ASICs exists for Stratix II devices.

1.1.6   Design flow

Design flow for Stratix II FPGAs is based on the Quartus II software for high-density FPGAs, which provides a comprehensive suite of synthesis, optimization, and verification tools in a single, unified design environment. Quartus II includes integrated development environment for Nios II embedded processors. Using the SOPC Builder design tool in the Quartus II software, designers select from the wide array of IP components, including memory, interface, control, and user-created functions, customize them for the particular application, and connect them – automatically generating hardware, software, and simulation models for the custom implementation.

1.1.7   Application area

STRATIX II FPGAs are very flexible allowing realization of different applications. Due to their high memory density Stratix II devices are an ideal choice for memory intensive applications. Using DSP blocks, Stratix II FPGAs can easily meet the DSP throughput requirements of emerging standards and protocols such as JPEG2000, MPEG-4, 802.11x, code-division multiple access 2000 (CDMA2000), HSDP and W-CDMA.

1.2   ALTERA Cyclone II

Cyclone II FPGAs [3] have been designed from the ground up for the lowest cost. The Cyclone II FPGA family offers a customer-defined feature set, high performance and low power consumption combined with high density. Altera claims that Cyclone II FPGAs offer the lowest cost per logic element among all commercially available devices and thus can support complex digital systems on a single chip at a cost that rivals that of ASICs.

1.2.1   Architecture

Cyclone II devices contain a two-dimensional row- and column-based architecture to implement custom logic. Column and row interconnects of varying speeds provide signal interconnects between logic array blocks (LABs), embedded memory blocks and embedded multipliers. The logic array consists of LABs, with 16 logic elements (LEs) in each LAB. A logic element (LE) is a small unit of logic providing efficient implementation of user logic functions. LABs are grouped into rows and columns across the device.

The smallest unit of logic in the Cyclone II architecture, the LE, is compact and provides advanced features with efficient logic utilization. Each LE features:

- A four-input look-up table (LUT), which is a function generator that can implement any function of four variables,
- a programmable register,
- a carry chain connection,
- a register chain connection
- and ability to drive all types of interconnects.

Each LE operates either in normal or in arithmetic mode (each one using LE resources differently). The architecture of LE is shown in Figure 3-2.

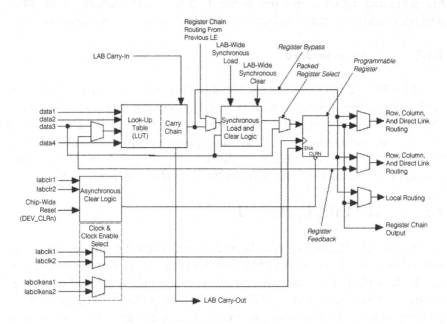

Figure 3-2. Cyclone II logic element structure

The Cyclone II embedded memory consists of columns of M4K memory blocks. The M4K memory blocks include input registers that synchronize writes and output registers to pipeline designs and improve system performance. Each M4K block can implement various types of memory with or without parity, including true dual-port, simple dual-port, and single-port

RAM, ROM, and first-in first-out (FIFO) buffers. Each M4K block has a size of 4,608 RAM bits.

Cyclone II devices have up to 150 embedded multiplier blocks optimized for multiplier-intensive digital signal processing (DSP) functions. Designers can use the embedded multiplier either as one 18-bit multiplier or as two independent 9-bit multipliers. Embedded multipliers can operate at up to 250 MHz (for the fastest speed grade) for 18×18 and 9×9 multiplications when using both input and output registers. Each Cyclone II device has one to three columns of embedded multipliers that efficiently implement multiplication functions.

Cyclone II devices support differential and single-ended I/O standards, including LVDS at data rates up to 805 megabits per second (Mbps) for the receiver and 622 Mbps for the transmitter, and 64-bit, 66-MHz PCI and PCI-X for interfacing with processors and ASSP and ASIC devices.

Cyclone II devices range in density from 4,608 to 68,416 LEs. Cyclone II devices offer between 119 to 1,152 Kbits of embedded memory with a maximum clock speed of 250 MHz. Cyclone II devices provide a global clock network and up to four phaselocked loops (PLLs). The global clock network consists of up to 16 global clock lines that drive throughout the entire device.

1.2.2   Granularity

Cyclone II architecture is a fine grain architecture with embedded hardwired word level modules.

1.2.3   Technology

Cyclone II devices are manufactured on 300-mm wafers using TSMC's 90-nm, 1.2-V, all-layer copper SRAM, low-k dielectric process technology, the same proven process used with Altera's Stratix II devices.

1.2.4   Reconfiguration

Cyclone II FPGAs are statically reconfigurable. Cyclone II devices are configured at system power-up with data stored in an Altera configuration device or provided by a system controller. Serial configuration allows configuration times of 100 ms. After a Cyclone II device has been configured, it can be reconfigured in-circuit by resetting the device and loading new configuration data.

1.2.5   Other issues

The Cyclone II FPGA family is fully supported by Altera's recently introduced Nios II family of soft processors. A Nios II design in a Cyclone II FPGA offers more than 100 DMIPs performance. With a Nios II processor, a designer can build a complete system on a programmable chip (SOPC) on any Cyclone II device, providing new alternatives to low- and mid-density ASICs.

1.2.6   Design flow

All Cyclone II devices are supported by the no-cost Quartus II Web Edition software. Quartus II software provides a comprehensive suite of synthesis, optimization and verification tools in a single, unified design environment. Designers can select from a large portfolio of intellectual property (IP) cores and download Altera's unique OpenCore Plus version of the chosen core(s). The Quartus II software is used to integrate and evaluate the cores in Cyclone II devices. Quartus II includes integrated development environment for Nios II embedded processors.

1.2.7   Application area

Cyclone II FPGAs are ideal for cost sensitive applications.

1.3   Xilinx Virtex 4

The Virtex-4 family [12] is the newest generation FPGA from Xilinx. Virtex-4 FPGAs include three families (platforms): LX, FX and SX. Choice and feature combinations are offered for all complex applications. The basic Virtex-4 building blocks are an enhancement of those found in the popular Virtex devices allowing upward compatibility of existing designs. Combining a wide variety of flexible features, the Virtex-4 family enhances programmable logic design capabilities and is a powerful alternative to ASIC technology.

1.3.1   Architecture

The configurable logic block (CLB) resource of Xilinx Virtex 4 is made up of four slices. Each slice is equivalent and contains: two function generators, two storage elements, arithmetic logic gates, large multiplexers, fast carry look-ahead chain and horizontal cascade chain. The function generators are configurable as 4-input look-up tables (LUTs). Two slices in a

CLB can have their LUTs configured as 16-bit shift registers, or as 16-bit distributed RAM. In addition, the two storage elements are either edge-triggered D-type flip-flops or level sensitive latches. Each CLB has internal fast interconnect and connects to a switch matrix to access general routing resources.

The general routing matrix (GRM) provides an array of routing switches between each component. Each programmable element is tied to a switch matrix, allowing multiple connections to the general routing matrix. The overall programmable interconnection is hierarchical and designed to support high-speed designs. All programmable elements, including the routing resources, are controlled by values stored in static memory cells. These values are loaded in the memory cells during configuration and can be reloaded to change the functions of the programmable elements.

The block RAM resources are 18 Kbit true dual-port RAM blocks, programmable from 16Kx1 to 512x36, in various depth and width configurations. Each port is totally synchronous and independent, offering three "read-during-write" modes. Block RAM is cascadable to implement large embedded storage blocks. Additionally, back-end pipeline registers, clock control circuitry, built-in FIFO support and byte write enable are new features supported in the Virtex-4 FPGA.

The Xtreme DSP slices contain a dedicated 18x18-bit 2's complement signed multiplier, adder logic and a 48-bit accumulator. Each multiplier or accumulator can be used independently. These blocks are designed to implement extremely efficient and high-speed DSP applications.

Most popular and leading-edge I/O standards (both single ended and differential) are supported by programmable I/O blocks (IOBs). In larger devices 10-bit, 200 kSPS analog-to-digital converter is included in the built-in system monitor block.

Additionally, FX devices support integrated hardwired high-speed serial transceivers that enable data rates up to 11.1 Gb/s per channel and 10/100/1000 Ethernet media-access control (EMAC) cores.

Virtex 4 FX devices support one or two hardwired IBM PowerPC 405 RISC CPUs (up to 450 MHz) with the auxiliary processor unit interface, which allows optimized FPGA based coprocessor connection. PowerPC 405 CPU is based on a 32-bit Harvard architecture with a five-stage execution pipeline supporting a CoreConnect bus architecture. Instruction and data L1 caches of 16 KB each are integrated.

Virtex 4 devices achieve clock rates of 500 MHz. Virtex 4 devices have logic densities of up to 200000 logic cells. Memory densities of up to 9935 kbits for block RAM and up to 1392 kbits distributed RAM are supported. DSP slices of up to 512 may be included leading to a 256 GMACs DSP performance.

1.3.2   Granularity

Virtex 4 architecture is a fine grain architecture with embedded hardwired word level modules and complete PowerPC CPUs.

1.3.3   Technology

Virtex-4 devices are produced on a state-of-the-art 90 nm triple oxide (for low power consumption) copper process, using 300 mm (12 inch) wafer technology. The core voltage of the devices is 1.2 V.

1.3.4   Reconfiguration

Virtex 4 FPGAs are dynamically (partially) reconfigurable devices.

1.3.5   Other issues

Optional 256-bit AES decryption is supported on-chip (with software bitstream encryption) providing Intellectual Property security.

1.3.6   Design flow

Xilinx ISE development system is used to map applications on the logic part of Virtex 4 devices. Advanced verification and real-time debugging is offered by ChipScope Pro tools. More than 200 pre-verified IP cores are available for Virtex 4 devices. The EDK PowerPC development kit is used for the realization of functionality on PowerPC CPUs.

1.3.7   Application area

Virtex-4 LX FPGAs are suitable for high-performance logic applications. Virtex-4 FX devices are well suited for high-performance, full-featured solution for embedded platform applications. Virtex-4 SX devices are a good solution for high-performance Digital Signal Processing (DSP) applications.

1.4   Xilinx Spartan-3

The Spartan-3 family of Field-Programmable Gate Arrays [10] is specifically designed to meet the needs of high volume, cost-sensitive consumer electronic applications. The Spartan-3 family builds on the success of the earlier Spartan-IIE family by increasing the amount of resources, the use of the state-of-the-art Virtex-II technology and the advanced process technology.

1.4.1 Architecture

Each Configurable Logic Block (CLB) comprises four interconnected slices, as shown in Figure 3-3. These slices are grouped in pairs. Each pair is organized as a column with an independent carry chain. All four slices have the following elements in common: two logic function generators, two storage elements, wide-function multiplexers, carry logic, and arithmetic gates. Both the left-hand and right-hand slice pairs use these elements to provide logic, arithmetic, and ROM functions. Besides these, the left-hand pair supports two additional functions: storing data using Distributed RAM and shifting data with 16-bit registers. The RAM-based function generator (Look-Up Table) is the main resource for implementing logic functions.

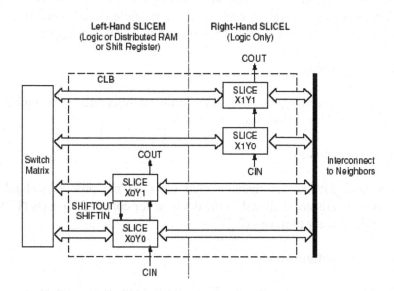

Figure 3-3. Spartan-3 CLB structure

Spartan-3 devices support block RAM, which is organized as configurable, synchronous 18Kbit blocks. Block RAM stores efficiently relatively large amounts of data. The aspect ratio – i.e., width vs. depth – of each block RAM is configurable. Furthermore, multiple blocks can be cascaded to create still wider and/or deeper memories. The blocks of RAM are equally distributed in 1 to 4 columns.

There are four kinds of interconnect: Long lines, Hex lines, Double lines, and Direct lines. Long lines connect to one out of every six CLBs; hex lines connect one out of every three CLBs; double lines connect to every other CLB. Direct lines afford any CLB direct access to neighboring CLBs.

Spartan-3 devices provide embedded multipliers that accept two 18-bit words as inputs to produce a 36-bit product. The input buses to the multiplier accept data in two's-complement form (either 18-bit signed or 17-bit unsigned). One such multiplier is matched to each block RAM on the die. The close physical proximity of the two ensures efficient data handling. Cascading multipliers permits multiplicands more than three in number as well as wider than 18-bits. Two multiplier versions are possible: one asynchronous and one with registered output.

Spartan-3 devices have logic densities of up to 74880 logic cells (corresponding to 5 million system gates). A system clock rate of up to 326 MHz is supported. Memory densities range from 72 to 1872 kbits of block RAM and 12 to 520 kbits of distributed RAM. The number of hardwired multipliers can be up to 104. Spartan devices include up to 784 I/O pins with 622 Mb/s data transfer rate per I/O. Seventeen single-ended signal standards and seven differential signal standards including LVDS are supported.

1.4.2   Granularity

Spartan-3 architecture is a fine grain architecture with embedded hardwired word level modules.

1.4.3   Technology

Spartan-3 FPGAs are manufactured on a 90 nm process technology. Three power rails are included in the devices: for core (1.2V), I/Os (1.2V to 3.3V) and auxiliary purposes (2.5V).

1.4.4   Reconfiguration

Spartan-3 FPGAs are dynamically (partially) reconfigurable devices.

1.4.5   Other issues

Spartan-3 devices allow integration of MicroBlaze soft processor, PCI, and other cores.

1.4.6   Design flow

Implementation of applications on Spartan-3 devices is fully supported by Xilinx ISE development system, which includes tools for synthesis, mapping, placement and routing. The EDK Microblaze development kit is used for the realization of functionality on Microblaze cores.

1.4.7   Application area

Because of their low cost, Spartan-3 FPGAs are ideally suited to a wide range of consumer electronics applications, including broadband access, home networking, display/projection and digital television equipment.

2.   INTEGRATED CIRCUIT DEVICES WITH EMBEDDED RECONFIGURABLE RESOURCES

Integrated circuits with embedded reconfigurable resources represent an alternative to FPGA ICs. These architectures are in principle based on a combination of a programmable CPU and a reconfigurable array of word level (coarse grain) data path units. Such devices mainly target DSP applications and are competitors of conventional DSP instruction set processors as well. The technology is less mature than FPGAs, however it promises important advantages over FPGAs such as power and silicon area efficiency. The major issue is the efficient compilation on the coarse grain reconfigurable resources.

2.1   ATMEL Field Programmable System Level Integrated Circuits (FPSLICs)

The Atmel's AT94 Series of Field Programmable System-Level Integrated Circuits (FPSLICs) [2] are combinations of the Atmel AT40K SRAM FPGAs and the Atmel AVR 8-bit RISC microcontroller with standard peripherals.

2.1.1   Architecture

The architecture of AT94K family is shown in Figure 3-4. The embedded AVR core is based on an enhanced, C code optimized, RISC architecture that combines a rich instruction set (more than 120 instructions) with 32 general-purpose working registers. All 32 registers are directly connected to the ALU, allowing two independent registers to be accessed in one single instruction executed in one cycle. AVR includes the full complement of peripherals such as SPI, UART, timer/counters and a hardware multiplier.

SRAM delivers one-cycle operation at up to 40 MHz, which translates into about 30 MIPS for the AVRs pipeline RISC design. For flexibility, the 36 KB of dynamically allocated AVR SRAM can be partitioned between x16 program store and x8 data RAM. For example, one setup might dedicate 20 and 16 KB for program and data respectively, another 32 and 4 KB.

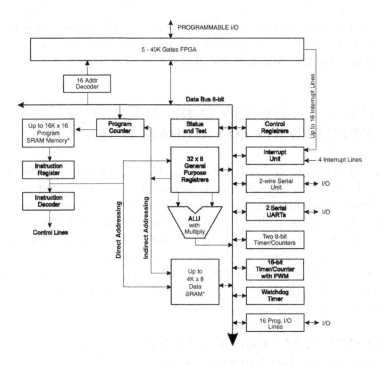

Figure 3-4. Atmel FPSLIC AT94K Architecture

The AVR core and FPGA connection is based on a simple approach that treats the FPGA much like another onboard 8-bit peripheral. There is an address decoder for generating up to 16 pseudochip selects into the FPGA and, going the other way, 16 interrupt lines that are fed from the FPGA into the AVR. The MCU has access to the FPGA's eight global clocks and can drive two of them relying on its own combination of internal and external oscillators, clock dividers, timer/counters and so on.

The FPGA core is based on a high-performance DSP optimized cell. FPSLIC devices include 5,000 to 40,000 gates of SRAM-based AT40K FPGA and 2 - 18.4 Kbits of distributed single/dual port FPGA user SRAM.

2.1.2   Granularity

The architecture of AT94 devices represents fine-grained architecture as far as programmable logic is concerned.

2.1.3   Technology

FPSLIC devices are fabricated on high-performance, low-power, 3.0V–3.6V, 0.35μ CMOS five-layer metal process.

2.1.4   Reconfiguration

The AT40K SRAM FPGA family is capable of implementing Cache Logic (Dynamic full/partial logic reconfiguration, without loss of data, on-the-fly) for building adaptive logic and systems. As new logic functions are required, they can be loaded into the logic cache without losing the data already there or disrupting the operation of the rest of the chip, replacing or complementing the active logic.

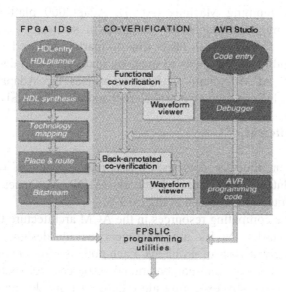

Figure 3-5. System Designer design flow

2.1.5   Design flow

Atmel provides System Designer tool suite (see Figure 3-5) that coordinates microcontroller and FPGA development with source-level debug and full hardware visibility. For implementation, the package includes place-and-route, floor planning, macro generators and bitstream utilities.

2.1.6   Application area

Atmel's AT94K series FPSLIC device provides the logic, processing, control, memory and I/O functions required for low-power, high-performance applications including among others: PDA and cell phone after-market products, GPS, portable test equipment, point-of-sale and security or wireless Internet appliances.

2.2   QuickSilver ADAPT2000 Adaptive Computing Machine System IC Platform

QuickSilver Technology Adapt2000 system platform [1], based on adaptive computing technology, attempts to integrate the silicon capability of ASIC, DSP, FPGA and microprocessor technologies within a single IC, an Adaptive Computing Machine (ACM). Adapt2000 platform aims at achieving custom-silicon capability designed in software – in weeks or months instead of years – with faster time to market, reduced development costs and the ability for designers to focus on innovating and developing IP. The Adapt2000 ACM system platform comprises the Adapt2400 ACM architecture, the InSpire Node Control Kernel and the InSpire SDK tool set.

2.2.1   Architecture

Adapt2400 architecture consists of two major types of components: Nodes and Matrix Interconnect Network (MIN). A generic view of Adapt2400 architecture is shown in Figure 3-6.

Nodes are the computing resources in the ACM architecture that perform the processing tasks. Nodes are heterogeneous by design, each being optimized for a given class of problems. Each node is self-contained with its own controller, memory, and computational resources. As such, a node is capable of independently executing algorithms that are downloaded in the form of binary files. Nodes are constructed of three basic components: The Node Wrapper, Nodal Memory and the Algorithmic Engine. The Node Wrapper has two major functions: a) to provide a common interface to the MIN for the heterogeneous Algorithmic Engines and b) to make available a common set of services associated with inter-node communication and task management. Each node is nominally equipped with 16 kilobytes of nodal memory organized as four 1k x 32 bit blocks. When building an ACM, memories can be adjusted in size, larger or smaller, to optimize cost or increase the flexibility of a specific node. Each heterogeneous node type is distinguished by its Algorithmic Engine. The computational resources of each node type are closely matched and optimized to satisfy a finite range of algorithms.

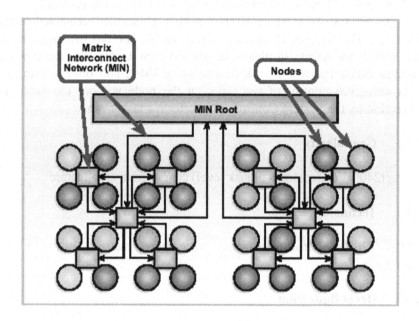

Figure 3-6. Generic view of Adapt2400 architecture

There are three classes of nodes in adaptive computing:

- *Adaptive nodes* support the heavy algorithmic elements that require complex control. They have a high degree of programmability and computational unit adaptability.
- *Domain nodes* are designed for the really complex pieces of the algorithms. Domain Nodes perform at speeds comparable to pure ASIC designs. Their control mechanisms are finite state machines.
- *Programmable nodes* are designed to support large code bases that do not demand much processing power. Designers are also able to build their own fully customized Algorithmic Engines and memory structures, and place them inside the Node Wrapper.

The Matrix Interconnect Network (MIN) ties the heterogeneous nodes together, and carries data, configuration binary files, and control information between ACM nodes, as well as between nodes and the outside world. This network is hierarchical in structure, providing high bandwidth between adjacent nodes for close coupling of related algorithms, while facilitating easy scaling of the ACM at low silicon overhead. Each connection between blocks within the MIN structure simultaneously supports 32 bits of data payload in each direction. Data within the MIN is transported in single 32-bit word packets, with addressing carried separately. Each 32-bit transfer within the MIN can be routed to any other node or external interface, with the MIN bandwidth fully shared between all the nodes in the system.

An Adapt2400 ACM has a built-in System Controller connected to the MIN Root. The System Controller is responsible for the management of tasks within an ACM. In this role, the System Controller sets up the individual Node Hardware Task Managers (HTMs), and once set up, the HTMs are given control of the tasks on the node without the need for intervention by the System Controller to perform a task swap.

2.2.2   Granularity

Adapt2400 architecture is a (task level) coarse grain architecture.

2.2.3   Technology

ADAPT2000 platform instances have been realized on 0.13 μm technologies.

2.2.4   Reconfiguration

Adapt2400 ACM architecture dynamically reconfigures during operation. ACM nodes are configured/programmed using a binary file (SilverWare), which is much smaller than that of a typical FPGA configuration file, and is comparable to the program size of a DSP or RISC processor. The smaller binary file size, combined with hardware specifically designed to adapt on the fly, allows the function of a node to change in as little as a few clock cycles.

2.2.5   Design flow

The Inspire SDK Tool Set by QuickSilver is a complete development system for the Adapt2400 ACM Architecture that provides a unified design environment that enables realization of an ACM within a single IC. The Inspire SDK comprises the SilverC development language (ANSI-C derivative), module linker, assembler for each node type and the InSpire Simulation Platform, including the ACM Verification SwitchBoard. The latter, provides multi-mode verification of ACM designs using any combination of the C Virtual Node (CVN), Inspire Simulation Platform,

InSpire Emulator, and an actual ACM device. The Inspire SDK is completely software-based and supports all phases of development, from high-level system simulation to compiled binaries running on an emulator or target IC. Its Adapt2400 SilverStream Design Flow enables developers to freely express system functionality without the need to consider hardware partitioning, task threading, or memory allocation. The InSpire SDK also enables engineers to create custom Adapt2400 architecture cores in simulation and assemble new nodal combinations for exploring a wide variety of ACM hardware configurations.

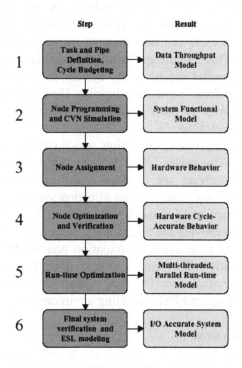

Figure 3-7. ACM design flow

The development flow for the Adapt2400 ACM Architecture is based on the use of a dataflow model of the system under development. In this methodology the system is represented in a series of top-down dataflow models that use successive refinement techniques to build up to a final hardware implementation. The ACM SilverStream Design Flow supports the task-based "execute when ready" asynchronous nature of the Adapt2400 ACM Architecture without requiring expert hardware knowledge on the part of the developer. The design flow consists of up to six steps as shown in Figure 3-7:

- The first step consists of: (a) modeling the dataflow of the system under development by using SilverC to define tasks, and pipes between the tasks, (b) assigning a cycle budget to each task and (c) simulating the data throughput of the system.
- The second step is to define the function of each task using ANSI-C, and then verifying the behavioral integrity of the system using C Virtual Nodes (CVN).
- The third step is node type and node instance assignment.
- The fourth step is hardware optimization with node verification using the node-type compilers or assemblers, and the appropriate node simulators. Step four provides an I/O accurate model of the system operation. Each node can be simulated using the ACM Verification SwitchBoard. This module in the InSpire Simulation Platform allows developers to model the hardware system as CVNs on the InSpire Adapt2400 Platform Emulator, InSpire Development Board, or a target device. Any of these models can be used in combination or individually at any time.
- The fifth step is run-time optimization, which consists of assignment of multiple tasks to nodes. The InSpire Simulation Platform and Performance Analyzer are used to determine which tasks can be assigned to the same node without affecting system operation. In this step, performance and hardware-size trade-offs can easily be made and analyzed to provide the best fit for system requirements.
- The sixth step is final system simulation and verification using the InSpire Simulation Platform to ensure overall system compliance with design specifications. The final system models contain SystemC APIs for inclusion into ESL modeling environments.

2.2.6   Application area

QuickSilver claims that ACM-enabled devices provide high performance, small silicon area, low power consumption, low cost and architecture flexibility and scalability – the ideal attributes for handheld, mobile and wireless products that span multiple generations. They particularly target signal and image processing applications.

2.3   IPflex DAPDNA-2 processor

The DAPDNA Dynamically Reconfigurable Processor [4] developed by IPFlex Inc. aims at providing "hardware performance" while maintaining "software flexibility.

2.3.1 Architecture

The DAPDNA-2 dynamically reconfigurable processor is a dual-core processor, comprised of IPFlex's own DAP high-performance RISC core, paired with the DNA two-dimensional processing matrix. The DAPDNA-2 processor can operate at 166 MHz. The DAP RISC core (32 bit with 8 kbytes data cache and 8 kbytes instruction cache) controls the processor's dynamic reconfiguration, while portions of an application that require high-speed processing are handled by the DNA matrix, which provides both parallel and pipelined processing. The DNA matrix is an array of 376 Processing Elements (PE) – comprised of computation units, memory, synchronizers, and counters. The total RAM of the DNA array is 576 kbytes. The DNA matrix circuitry can be reconfigured freely into the structure that is most optimal for meeting the needs of the application in demand. One foreground and three background banks are available on-chip to store different configurations. Additional banks can be loaded from external memory on demand. The architecture of DAPDNA-2 processor is shown in Figure 3-8.

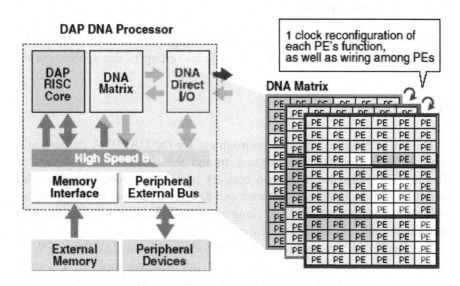

Figure 3-8. DAPDNA-2 processor architecture

Large on-chip memory reduces the need to access off-chip memory a process that often becomes a performance bottleneck. This feature allows the DNA to provide the maximum possible parallel processing performance. Since the memory is distributed throughout the processing array, there is plenty of available memory bandwidth.

The DAPDNA-2 has six channels of DNA Direct I/O, which provides the interface for transfering data directly onto or out of the DNA matrix. Each channel of DNA Direct I/O is 32-bit wide and operates at the maximum DAPDNA-2 system clock frequency of 166 MHz. The DNA Direct I/O can be also used to communicate directly with external devices, bringing data in for processing on the DNA matrix, bypassing the Bus Switch and memory interface.

2.3.2   Granularity

The DNA matrix architecture is a coarse grain reconfigurable architecture.

2.3.3   Technology

The DAPDNA-2 processor comes in a 156-pin FCBGA package. The power supply for the core is 1.2 V while for the I/Os is 2.5 V.

2.3.4   Reconfiguration

DAPDNA processor is dynamically reconfigurable and can change its hardware configuration in one clock cycle according to the application on demand.

2.3.5   Design flow

The integrated development environment for the DAPDNA dynamically reconfigurable processor is designed around the concept of "Software to Silicon". The Software to Silicon concept means that even someone who doesn't know how to design hardware can develop a product by designing an application using a high-level language, and having that application seamlessly implemented as a hardware.

The DAPDNA processor series is provided with the DAPDNA-FW II Integrated Development Environment, a full-featured tool set covering everything from algorithm design to validation of an application running on the actual hardware. DAPDNA-FW II provides compilers for algorithms written in MATLAB/Simulink and C with data flow extension.

DAPDNA-FW II environment supports three different design methodologies, giving the designer the flexibility to choose the most appropriate design method. The first option is to use the Data Flow C (DFC) Compiler. In this case it is possible to use the C programming language to directly create code for the dynamically reconfigurable processor. In a development process built around the DFC compiler, the designer can create code directly using the C programming language, which reduces the development time. The second option is to use the DNA Blockset, which allows algorithm design and verification using MATLAB, Simulink (from The MathWorks Inc). DNA Blockset enables a seamless design flow from algorithm design to implementation in the DAPDNA-2 processor, all within the MATLAB/ Simulink environment. The third option is the DNA designer which is a GUI-based development environment allowing the designer to drag-and-drop representations of the DAPDNA Processing Elements (PEs), supporting graphical construction of processing algorithms.

2.3.6   Application area

IPflex claims that the DAPDNA-2 is the world's first general-purpose dynamically reconfigurable processor. It is suitable for applications that demand, high performance and support for a wide range of processing tasks. It also provides a solution that is optimal for today's marketplace, with its demand for short-run, mixed-model production. Target applications include industrial performance image processing (for factory automation, inspection systems), broadcast and medical equipment, high precision high speed image processing (multi-function peripherals, laser printers etc), base stations (cellular, PHS, etc), accelerators for image processing, data processing and technical computation, security equipment, encryption accelerators and software defined radio.

2.4   Motorola MRC6011 Reconfigurable fabric device

The MRC6011 device is the first reconfigurable compute fabric (RCF) device from Freescale Semiconductor [7]. It is a highly integrated system on a chip (SoC) that combines six reconfigurable compute fabric (RCF) cores into a homogeneous compute node. The programmable MRC6011 device aims at offering system-level flexibility and scalability similar to a programmable DSP while achieving the cost, power consumption and processing capability of a traditional ASIC-based approach.

2.4.1   Architecture

The MRC6011 RCF cores are accessible in two scalable modules, each containing three RCF cores, via two multiplexed data input (MDI) interfaces and two slave I/O Interfaces. Each MDI interface can communicate with up to 12 channels (antennas for example), and each RC controller can manipulate the data from two channels. The data processed by the RCF cores goes either to one of the two slave I/O bus interfaces (compatible with industry-wide DSPs) or to another core within the same module or the adjacent module. External interfaces include the MDI interfaces and slave I/O bus interfaces (supporting DSP bootstrapping) operating at up to 100 MHz, and a JTAG port for real-time debugging. The architecture of the MRC6011 device is shown in Figure 3-9.

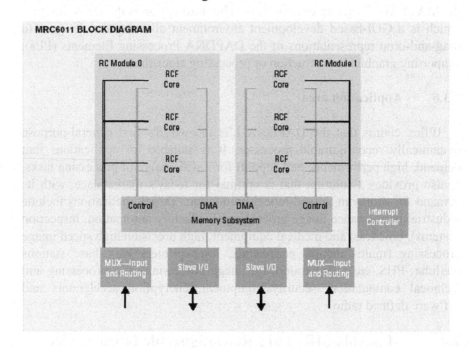

Figure 3-9. Architecture of MRC6011 device

Each RCF core includes an optimized 32-bit RISC processor (allowing efficient C code compilation) with instruction (4 kbytes) and data caches (4 kbytes). The reconfigurable computing (RC) array includes 16 reconfigurable processing units with 16 bit data paths including a pipelined MAC unit. The RCF core also includes a two-channel input buffer (8 kbytes), a large frame buffer (40 kbytes) with eight address generation units (AGUs), a special-purpose complex correlation unit supports spreading, complex scrambling, complex correlation on 8-bit and 4-bit samples and a single and burst transfer DMA controller.

At 250 MHz, the six-core MRC6011 device delivers a peak performance of 24.0 Giga complex correlations per second with a sample resolution of 8 bits for I and Q inputs each, or even 48.0 Giga complex correlations per second at 4-bit resolution.

2.4.2   Granularity

The architecture of the MRC6011 is a coarse grain architecture based on the word level reconfigurable data paths of the RC arrays.

2.4.3   Technology

MRC6011 devices are manufactured on a 0.13 μm process technology. The internal logic voltage is 1.2 V while the input/output voltage is 3.3 V. The core maximum operating frequency is 250 MHz while the maximum operating frequency for all off-core buses is 100 MHz.

2.4.4   Reconfiguration

MRC6011 is a dynamically reconfigurable multi-context device.

2.4.5   Design flow

Design flow for MRC6011 is based on C and assembly programming. The CodeWarrior Development Studio for Freescale RCF Baseband Signal Processors is a complete development environment for Freescale Reconfigurable Compute Fabric (RCF) based devices. The CodeWarrior Development Studio is a complete code development studio and includes: a) the Project Manager that provides anything required for configuring and managing complex projects, b) the Editor and Code Navigation System that allows creation and modificaton of source code and c) the graphical level debuggers.

CodeWarrior Development Studio, in concert with the PowerTAP Pro hardware target interface, provides a multi-core debugging environment that allows for quick single stepping as well as fast downloads of very large target files. In case of multiple MRC6011 products, it is possible to connect the JTAG connections in a way allowing talking to any of the MRC6011's through a single PowerTAP device. Since PowerTAP has Ethernet as it's connection method to CodeWarrior, debugging can be done remotely as well as providing a mechanism to share a single resource among several engineers. Functional testing effort can be minimized through utilization of CodeWarrior Development Studio's full scripting capability.

2.4.6   Application area

Highly flexible and programmable, the MRC6011 processor provides an efficient solution for computationally intensive applications, such as wideband code division multiple access (WCDMA), CDMA2000 and TD-SCDMA baseband processing, including chip rate, symbol rate and advanced 3G functions such as adaptive antenna (AA) and multi-user detection (MUD).

2.5   picoChip PC102 – picoArray processor

The PC102 is the 2nd generation of the picoArray highly parallel processing architecture developed by picoChip [9]. The picoChip's PC102 picoArray processor is a signal processing device optimised for next generation wireless infrastructure. The solution can be described as a "Software System on Chip" (SSoC): fast enough to replace FPGAs or ASICs but with the flexibility and ease of programming of a processor. PC102 picoArray processor offers scalability allowing extremely large systems to be built by connecting dozens of processors.

2.5.1   Architecture

The architecture emphasises ease of design/verification and deterministic performance for embedded signal processing – especially wireless. The picoArray combines hundreds of array elements, each with a versatile 16 bit RISC processor (3 way LIW with Harvard architecture) with local data and program memory connected by a high-speed interconnect fabric. The architecture is heterogeneous with four types of element optimised for different tasks such as DSP or wireless specific functions. As well as the standard array elements, others handle control functions, memory intensive and DSP-oriented operations. Multiple array elements can be programmed together as a group to perform particular functions ranging from fast processing such as filters and correlators, through to the most complex control tasks.

Within the picoArray core, array elements are organised in a two dimensional grid, and communicate over a network of 32 bit buses (the picoBus) and programmable bus switches. Array elements are connected to the picoBus by ports. The ports act as nodes on the picoBus and provide a simple interface to the bus based on *put* and *get* instructions in the instruction set. The inter-processor communication protocol is based on a time division multiplexing (TDM) scheme, where data transfers between processor ports occur during time slots, scheduled in software, and controlled using the bus switches. The bus switch programming and the scheduling of data transfers is fixed at compile time.

Around the picoArrray core are system interface peripherals including a host interface and an SRAM interface. Four high speed I/O interfaces connect to external systems or link picoArray devices together to build scalable systems. The basic concept of picoArray architecture is shown in Figure 3-10.

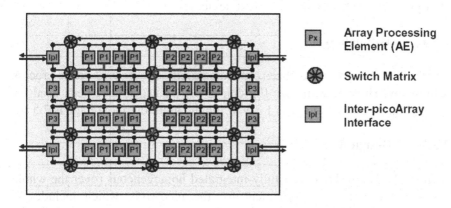

Figure 3-10. Basic concept of picoArray architecture

PC102 picoArray has huge processing resources for compute intensive datapath. It also has enormous amounts of general-purpose MIPS to handle the ever more complex control operations. The PC102 uses 348 array elements running at 160MHz, and with peak use can handle over 197,100 million instructions per second (MIPS), 147,800 million operations per second (MOPS) or 38,400 million multiply accumulate (MMAC) instructions per second – over 10 times the performance of other programmable solutions. The microprocessor interface is used to configure the PC102 device and to transfer data to and from the PC102 device using either a register transfer method or a DMA mechanism. The interface has a number of ports mapped into the external microprocessor memory area. Two ports are connected to the configuration bus within the PC102 and the others are connected to the picoBus. These enable the external microprocessor to communicate with the array elements using signals. Alternatively, the PC102 can self-configure (or boot) in standalone mode from a supported memory.

2.5.2   Granularity

PC102 processor's picoArray architecture is a (CPU level) coarse grain reconfigurable architecture based on 16 bit CPUs.

2.5.3   Reconfiguration

The picoArray architecture is totally programmable and can be configured at run time (single context device).

2.5.4   Technology

PC102 devices have been manufactured on a 0.13 μm process technology. High performance flip chip BGA packages have been used for packaging. The core voltage is 1.2 V while the input/output voltage is 2.5 V.

2.5.5   Design flow

picoChip's picoTools is a fully-integrated homogeneous (over the whole system) development environment for the picoArray which includes C compiler, assembler, debugger and cycle-accurate simulator, in which system performance is guaranteed by design (with complete predictability). picoChip also supplies a Library of Example Designs and a range of Development platforms.

The developer defines the structure and relationships between processes, completely specifying signal flows and timings. The individual processors are then programmed in standard C or assembler as blocks to be embedded within the structure. The entire design (structure, data-path and control) is debugged at the source level. This allows engineers to work on the whole system in an integrated way, rather than having to debug different technologies separately. The programming of the array is completely automatic, and the designer is abstracted from this implementation details. The output is a hardware configuration file containing the design and the timing information to run in the simulation. This creates a seamless "closed loop" flow from the simulator to the development kit through to the system. The picoChip architecture is extremely scalable, and applications can be run across multiple linked devices. The tools allow large designs to be simulated, placed and verified as easily as small ones. The architecture gives high levels of confidence in using multiple pre-verified blocks in a series of static software architectures that can be implemented at different times on the same hardware to give a truly reconfigurable system.

2.5.6   Application area

The PC102 is a communications processor, optimized for high capacity wireless digital signal processing applications. The device enables all layer 1 (physical layer) signal processing and layer 1 control to be implemented in software. The device is able to run any wireless protocols including WCDMA (FDD and TDD), cdma2000 and TD-SCDMA, or emerging standards such as 802.16 (WiMAX).

2.6 Leopard Logic Gladiator Configurable Logic Device

The Gladiator configurable logic device (CLD) [6] family represents the only digital logic device that combines Field Programmable Gate Array (FPGA) technology with hardwired Application Specific Integrated Circuit (ASIC) logic. Gladiator CLD aims at achieving much lower NRE charges than ASICs in combination with dramatically lower unit cost than complex FPGAs.

In its first steps Leopard Logic provided embedded FPGA IP cores for ASIC/SoC and foundry suppliers but industry's interest with respect to this approach was limited. Then Leopard Logic reinvented itself as a silicon supplier.

2.6.1 Architecture

The architecture of Gladiator CLD is shown in Figure 3-11. The basic building blocks of Gladiator CLD are the HyperBlox FP (Field Programmable) and the MP (Mask Programmable) fabrics, which are combined with optimized memories, Multiply-Accumulate units (MACs) and flexible high-speed I/Os.

Gladiator CLD is available in densities ranging from 1.6M up to 25M system gates with up to 10 Mbits of embedded memory. It supports system speeds up to 500MHz. Gladiator CLD includes high speed MAC units for fast arithmetic and DSP, up to 16 PLL controlled clock domains with frequency synthesis and division and, up to 16 DLL for phase shifting to support interface timing adjustment. Gladiator CLD offers flexible I/O options and supports several general purpose I/O standards. Gladiator CLD also supports DDR/QDR.

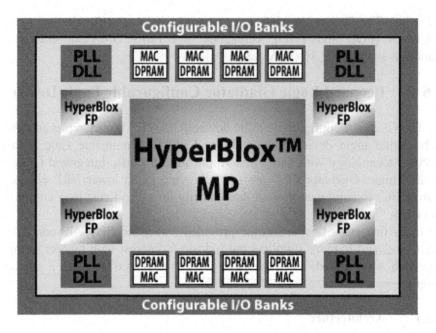

Figure 3-11. Architecture of Gladiator CLD

2.6.2   Granularity

The architecture of Gladiator CLD represents a fine grain architecture.

2.6.3   Technology

The HyperBlox FP fabric is based on Leopard Logic's proprietary HyperRoute FPGA technology that utilizes the industry's first fully hierarchical, multiplexer-based, point-to-point interconnect. This technology enables superior speed, utilization, predictability and reliability compared to legacy FPGA architectures. The HyperBlox MP fabric uses the same logic core cell architecture as HyperBlox FP but replaces the SRAM configuration with a single-layer via-mask configuration, called HyperVia. This technology provides significantly higher density, as well as increased performance and lower power.

2.6.4   Reconfiguration

The Gladiator CLD is statically field-upgradeable through embedded SRAM-based FPGA.

2.6.5 Design flow

The Gladiator CLD design flow is based on leading industry standard design tools and flows combined with Leopard Logic's highly optimized ToolBlox back end tools. Partitioning between the HyperBlox MP and FP sections of the device is done intuitively. Fixed and stable blocks of the design are mapped into the HyperBlox MP fabric, while high-risk blocks that are still in flux are mapped into the FP fabric. Designs are quickly and easily synthesized from RTL into a CLD device. Full timing closure is achieved based on accurate timing extraction performed by the user. Bitstreams for the FPGA sections of the device are generated automatically and can be downloaded into the device instantly. Partitioning between hard (MP) and soft (FP) functions is a snap with the ToolBlox design flow and the unified hardware architecture allows the allocation of design blocks even post-synthesis.

Starting from pre-processed wafers, users can implement substantial amounts of high speed logic in the mask-programmable (MP) section of the device. After sending the generated configuration data to Leopard Logic, first samples are delivered within weeks. This process is referred to as "marketization" because it transforms the generic device into a user or market segment specific device. Due to minimum mask and processing requirements, the Non-Recurring Engineering (NRE) costs for this process are an order of magnitude lower than for a traditional cell-based ASIC.

The "marketized" devices can be further customized and differentiated by programming the HyperBlox FP fabric. Like any other SRAM-based FPGA, this fabric allows for an unlimited number of reconfigurations by simply downloading a new bistream into the device, thus offering optimal in-field programmability.

2.6.6 Application area

Gladiator Configurable Logic Device is suitable for areas that today use a combination of Application Specific Standard Product (ASSP)/ASIC with standalone FPGAs such as networking (edge, access, aggregation, framers, communications controllers, backplane interfaces), storage (bridges, controllers, interfaces, glue logic) and wireless (DSP acceleration, chip rate processing, smart antenna, bridges, backplanes, glue logic). Across all markets, Gladiator is an ideal fit for the fast and cost-effective implementation of flexible format converters, protocol bridges, bus interfaces and glue logic functions.

3.   EMBEDDED RECONFIGURABLE CORES

As the System-on-Chip (SoC) world began to develop at the end of the 1990s, it was recognised that, to make the devices more useful, some form of programmable fabric would be needed. ASIC developers also considered embedded reconfigurable logic as one way to bring some form of field programmability to an otherwise dedicated product. The industry responded in an enthusiastic fashion and a number of reconfigurable hardware cores that can be embedded in SoCs/ASICs have been proposed since late 1990s. Two major architectures have been mainly considered: embedded FPGAs (fine grain) and reconfigurable arrays of word level data paths (coarse grain). Despite the initial enthusiasm several of these attempts failed commercially (Adaptive Silicon disappeared while Actel stopped their embedded FPGA technology activities). Major reasons were the high silicon area (it could require half the chip area to put a decent amount of programmable logic on it), and the power overheads of embedded FPGAs and the immature compilation techniques for the coarse grain reconfigurable arrays.

In October 2004 during the EDA Tech Forum in San Jose, it was projected that until the first quarter of 2005 two embedded FPGA cores for ASICs/SoCs will be put on the market - one by a combination of IBM and Xilinx and the other by STMicroelectronics. The major reason that could lead these attempts to commercial success is the use of 90 nm technologies.

3.1   Morpho Technologies MS1 Reconfigurable DSP cores

Morpho technologies reconfigurable DSP (rDSP) cores MS1-16 and MS1-64 [8] aim at providing hardware flexibility in implementing multiple applications, minimized levels of obsolescence, and low power consumption while lowering hardware costs. The cores are available as is, or may be custom designed and/or quickly integrated into any SoC, to fit the needs of the customer and application(s).

3.1.1   Architecture

The MS1 family of rDSPs is fully autonomous IP (soft, firm or hard) cores that function as co-processors to a host processor in a system. The MS1 rDSP architecture consists of a 32-bit RISC with 5 pipeline stages and built-in direct-mapped data and instruction cache, an RC Array with 8 to 64 Reconfigurable Cells (each having an ALU, MAC and optional complex correlator unit), Context memory with 32 to 512 context planes, a Frame Buffer with up to 2048 Kbytes in size, and three optional blocks specific to

3G-WCDMA base station applications (namely a Sequence Generator), an Interleaver and an IQ Buffer (16 bytes to 4Kbytes per antenna). A multi-master 128-bit DMA bus controller supporting burst transfers with both synchronous and asynchronous memory interface is also included in the MS1 architecture. The architecture of the RC array is shown in Figure 3-12.

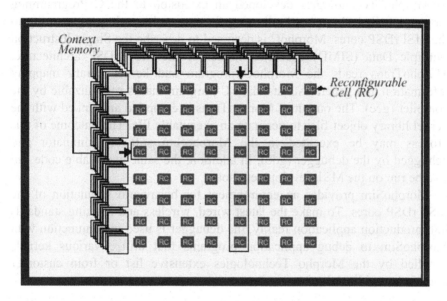

Figure 3-12. Architecture of Reconfigurable Cells array

3.1.2   Granularity

The reconfigurable cells array (RC) of Morpho technologies rDSP cores is a reconfigurable array of coarse grain data paths.

3.1.3   Technology

Evaluation devices are available in 0.18 µm and 0.13 µm process technologies with core voltages at 1.8V/1.2V and 3.3V digital I/O voltage.

3.1.4   Reconfiguration

Morpho technologies reconfigurable DSP (rDSP) cores are dynamically reconfigurable and can adapt on the fly to realize different applications. Switching from one application specific set of instructions to another is done on a single clock cycle.

3.1.5   Design flow

The MS1 rDSP cores and associated evaluation devices are accompanied with a complete tool chain that includes software development tools such as a compiler and translator, a simulator and a debug tool.

Morpho Technologies developed an extension to the C Programming language called "MorphoC" allowing for fast and simple programming to the MSI rDSP cores. MorphoC is designed to describe the Single Instruction Multiple Data (SIMD) execution model of the MS1 rDSP architecture. MorphoTrans reads the MorphoC program and kernel library mapping information and generates a standard C program that is recognizable by the compiler (gcc). The output of MorphoTrans is compiled and linked with the kernel library object files to generate an executable file. The outcome of this process may be executed in the MorphoSim software simulator and debugged by the debugger (gdb). In addition, the same executable code can also be run on the MS1 development board.

MorphoSim provides an environment for behavioral simulation of the MS1 rDSP cores. To make the latest wired, wireless and imaging standards into production application reality, the debugger is used in conjunction with MorphoSim to debug application programs that utilize various kernels supplied by the Morpho Technologies extensive list or from customer specific kernel libraries.

3.1.6   Application area

Morpho technologies reconfigurable DSP cores are capable of implementing the baseband processing of air interfaces such as WCDMA in addition to source processing such as MPEG4 and vocoders. In general Morpho technologies reconfigurable DSP cores are suitable for signal processing based products including communications equipment for wireless and wireline terminals and infrastructure, home entertainment and computer graphics/image processing.

3.2   PACT XPP IP cores

A PACT XPP processor or coprocessor [13] can be integrated in a System-on-Chip (SoC) and can be designed from a small set of macro blocks of which the largest is in the range of 90 kgates. The homogeneous architecture of XPP allows synthesizing each of the blocks separately and, in the second step, arranging the synthesized blocks hierarchically to the final array.

3.2.1 Architecture

An array of configurable processing elements is the heart of the XPP. The array is built from a very small number of different processing elements (PEs). ALU-PEs perform the basic computations. RAM-PEs are used for storage of data. The I/O elements connect the internal elements to external RAMs or data ports. The configuration manager loads programs onto the array. The architecture of the array is shown in Figure 3-13.

The ALU is a two input – two output ALU providing typical DSP functions such as multiplication, addition, comparison, sort, shift and boolean. All operations are performed within one clock cycle. The ALU can be utilized for addition, barrel shift and normalization tasks. The Forward Register is a specialized ALU that provides data stream control such as multiplexing and swapping. It introduces always one cycle pipeline delay.

The Communication Network allows point to point and point to multipoint connections from outputs to inputs of other elements. Up to 8 data channels are available for each horizontal direction. Switches at the end of the lines can connect the channel to the channel of the neighboring element.

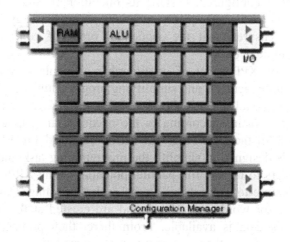

Figure 3-13. Architecture of XPP's array of configurable processing elements

The RAM Elements are arranged at the edges of the array and are nearly identical to the ALU PEs, however the ALU is replaced by a memory. The dual ported RAM has two separate ports for independent read and write operations. The RAM can be configured to FIFO mode (no address inputs needed) or RAM with 9 or more address inputs. The IP model allows to define the storage capacity. Typical values range from 512 to 2 k words.

Back Register and Forward Register can be configured to build a linear address generator. Thereby DMA to or from RAM can be done within one RAM-PE. Several RAM-PEs can be combined to a larger RAM with a contiguous address space.

I/O Elements are connected to horizontal channels. The standard I/O-Element provides two modes:

- *Streaming* Two ports per I/O Elements are configured to input or output mode. The XPP Packet handling is performed by a Ready-Acknowledge handshake protocol. Thus external data streams (e.g. from a A/D-converter) must not be synchronous to the XPP clock.
- *RAM* One output provides the addresses to the external RAM, the other is the bi-directional data port. External Synchronous Static RAMs are directly connected to the address ports, data ports and control signals. The maximum size of external RAMs depends on the data bus width of the XPP (e.g. 16 Mwords for the 24-bit architecture).

The Configuration Manager (CM) microcontroller handles all configuration tasks of the array. Initially it reads configurations through an external interface directly from S-RAMs into its internal cache. Then it loads the configuration (i.e. opcodes, routing channels and constants) to the array. As soon as a PE is configured, it starts its operation if data is available. Further on, the CM loads subsequent configurations to the array. The local operating system ensures, that the sequential order of configuration is maintained without deadlocks.

The structure of XPP array of configurable elements is very simple making the array homogeneous and simplifying programming and placing of algorithms. The IP model of XPP allows defining the size and arrangement of the processing elements according to the needs of the applications. In addition, the width of the Data Paths and ALUs can be defined between 8 and 32 bit. XPP is designed to simplify the programming task and to allow high level compilers to tap the full parallel potential of the XPP. The most important XPP feature to support this, is the packet handling. Data packets contain one processor word (e.g. 24-bit) and are created at the outputs of objects as soon as data is available. From there, they propagate to the connected inputs. If more then one input is connected to the output, the packet is duplicated. On the other hand, an XPP object starts its calculation only when all required input packets are available. If a packet can not be processed, the pipeline stalls until the packet is processed. This mechanism ensures correct operation of the algorithm under all circumstances and, the programmer does not need to care about pipeline delays in the array and how to synchronize to asynchronous external data streams.

3.2.2   Granularity

PACT XPP arrays architecture is a coarse grain reconfigurable architecture.

3.2.3   Technology

XPP cores are technology independent. PACT provides XPP cores as synthesizable Verilog RTL code.

3.2.4   Reconfiguration

XPP arrays allow fast dynamic reconfiguration. In contrast to FPGAs, XPP needs only Kbits for a full configuration; internal RAMs buffer data between the configurations. For optimal performance the number of data, which is calculated in one configuration, should be as high as possible to minimize the effect of the reconfiguration latency. Small parts of the array can be reconfigured without the need to stop calculations of other configurations on the same array.

3.2.5   Design flow

The XDS development suite supports co-development and co-simulation of systems with the XPP-array. The XDS is a complete set of tools for application development. Since in most applications XPP is used as a coprocessor to micro-controllers, the XDS provides seamless design-flow for both, the micro-controller and the XPP.

Derived from a data flow graph, algorithms are directly mapped onto the array. The Graphs's nodes define directly the functionality and operation of the ALU or other elements, whereas the edges define the connections between the elements. Such a configuration remains statically on the array and a set of data packets flows through this net of operators.

Applications are written in C or C++. In an environment with a micro-controller and the XPP as coprocessor, the software tasks are divided into two sections. The control-flow tasks are processed with the standard tools for the micro-controller and the high bandwidth data-flow tasks, that need support by the XPP, are compiled by the XPP-VC. This vectorizing C-compiler maps a subset of C to the XPP, and allows integrating optimized modules. These modules originate from a library, or are written for the application in the Native Mapping Language, NML. API functions for loading and starting of configurations, configuration sequencing, data exchange via DMA and task synchronization provide a comfortable environment for C-programmers who are familiar with embedded designs. The linker combines code of both sections, which can either be simulated by software, or uploaded to the target hardware. The integrated debugging tool for the micro-controller and the XPP, allows interactive test and verification of the simulation results or the hardware. The configuration and the dataflow in the XPP are visualized in a graphical tool.

3.3   Elixent DFA1000

The Elixent DFA1000 accelerator [5] was designed from the ground up to deliver on the promise of Reconfigurable Signal Processing (RSP). Utilizing the advanced D-Fabrix processing array. It aims at delivering huge benefits in performance, power consumption and silicon area. These attributes make it ideal for integration with RISC processors in mobile/consumer/communications applications that need the ultimate in signal or media processing. These advantages are delivered through silicon reuse. The DFA1000 accelerator implements "virtual hardware" – hardware accelerators for specific algorithms, implemented as simple configurations on the D-Fabrix processing array. When one algorithm completes, a new "virtual hardware" accelerator is loaded, performing the next task in the system's dataflow.

3.3.1   Architecture

The basis for Elixent's DFA1000 is the D-Fabrix processing array – a platform that realises the potential of Reconfigurable Algorithm Processing. The structure of D-Fabrix is simple – the components are 4-bit ALUs, registers and the "switchbox". Two of each are combined into a building block, the "tile". Hundreds or thousands of tiles are combined to create the D-Fabrix array. Special functions can be distributed through the array – for example, memory is always distributed to give fast, local storage with massive bandwidth. Creating wider execution units is simply a matter of combining ALUs – typically into 8, 12 or 16-bit units, but occasionally into far larger units. Much of the task of linking the ALUs together in this way is performed by the array's routing switchboxes. The architecture of the D-Fabrix array is shown in Figure 3-14.

The DFA1000 accelerator integrates several banks of local high-speed RAM next to the array. These are for often-used data; for example, they may be used as image linestores, or as audio buffers. These RAMs eliminate many high bandwidth accesses off-chip, improving power consumption while at the same time enhancing performance.

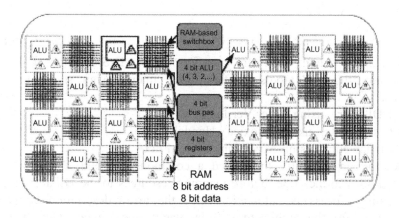

Figure 3-14. Architecture of D-Fabrix array

The DFA1000 also includes a peripheral set to facilitate its integration into SOC designs. The architecture offers high-speed data interfaces to the D-Fabrix core array. This allows high-speed data to be driven into the array directly, with low latency and no overhead on the system bus. These high-speed data interfaces are supplemented by the AMBA bus interface, used for programming the array, and transferring data to and from the host processor. This is typically a much lower bandwidth control and configuration path. The architecture also integrates local high-speed RAMs, directly accessible by the array or by the RISC; and of course the D-Fabrix array itself.

3.3.2   Granularity

DFA1000 architecture is a medium granularity architecture based on 4-bit ALUs.

3.3.3   Technology

DFA1000 will be made available in different industry standard processes. First realization was on a 0.18 µm technology.

3.3.4   Reconfiguration

DFA1000 can be dynamically reconfigured in microseconds.

3.3.5   Design flow

The key to using the DFA1000 accelerator is creating the high-performance virtual hardware configurations. D-Sign, the D-Fabrix algorithm processor's toolset offers, three main design styles for this purpose:

- HDL entry, using either Verilog or VHDL
- C-style entry, using Celoxica's Handel-C
- Matlab entry, using Accelchip's Accel-FPGA

All the design entry tools feed a common back-end. This performs optimisations to the code, before mapping resources to the D-Fabrix array. The entire process is automatic. Once the array description has been "compiled" for the architecture, it is placed and routed. This stage is analogous to the resource allocation phases that a compiler uses for a VLIW processor, allocating array resource to the functions within the algorithm. The output of the "place and route" tool is the final program.

3.3.6   Application area

D-Fabrix is suitable for several applications from networked multimedia (MPEG-4, JPEG, camera, graphics, rendering) to wireless (3G, CDMA, OFDM etc) or even security (RSA, DES, AES...).

REFERENCES

1. Adapt2000 QuickSilver Technologies (2004) Available at: http://www.qstech.com/default.htm
2. AT40K Atmel (2004) Available at: http://www.atmel.com/atmel/products/ prod39.htm
3. Cyclone II Altera (2004) Available at: http://www.altera.com/products/devices/cyclone2/cy2-index.jsp
4. DAPDNA IPFlex Inc (2004) Available at: http://www.ipflex.com/en
5. DFA1000 Elixent (2004) Available at http://www.elixent.com/products
6. Gladiator Leopard Logic (2004) Available at: http://www.leopardlogic.com/products/index.php
7. MRC6011 Freescale (2004) Available at: http://www.freescale.com/webapp/sps/site/prod_summary.jsp?code=MRC6011&nodeId=01279LCWs
8. MS1 Morpho Technologies (2004) Availabel at : http://www.morphotech.com/
9. picoArray picoChip (2004) Available at: http://www.picochip.com/technology/picoarray
10. Spartan-3 Xilinx (2004) Available at: http://www.xilinx.com/xlnx/xil_pro-dcat_landingpage.jsp?title=Spartan-3
11. Stratix II Altera (2004) Available at: http://www.altera.com/products/devices/stratix2/st2-index.jsp
12. Virtex-4 Xilinx (2004) Available at: http://www.xilinx.com/xlnx/xil_prodcat_landingpage.jsp?title=Virtex-4

13. XPP IP cores PACT (2004) Available at: http://www.pactcorp.com/

PART B

SYSTEM LEVEL DESIGN METHODOLOGY

Chapter 4

DESIGN FLOW FOR RECONFIGURABLE SYSTEMS-ON-CHIP

Konstantinos Masselos[1,2] and Nikolaos S. Voros[1]
[1] INTRACOM S.A., Hellenic Telecommunications and Electronics Industry, Greece
[2] Currently with Imperial College of Science Technology and Medicine, United Kingdom

Abstract: A top down design flow for heterogeneous reconfigurable Systems-on-Chip is
              presented in this chapter. The design flow covers issues related to system level
              design down to back end technology dependent design stages. Emphasis is
              given on issues related to reconfiguration, especially in system level where
              existing flows do not cover such aspects.

Key words: Design flow, system level, reconfiguration, reconfigurable Systems-on-Chip

1.   INTRODUCTION

Heterogeneous Systems-on-Chip (SoCs) with embedded reconfigurable resources form an interesting option for the implementation of wireless communications and multimedia systems. This is because they offer the advantages of reconfigurable hardware combined with the advantages of other architectural styles such as general purpose instruction set processors and application specific integrated circuits (ASICs). Furthermore, such SoCs allow customization on the way reconfigurable resources can be used (type and density of resources) depending on the targeted application or set of applications.

A generic view of a heterogeneous reconfigurable System-on-Chip is shown in Figure 4-1. Such a SoC will normally include instruction set processors (general purpose, DSPs, ASIPs), custom hardware blocks (ASICs) and reconfigurable hardware blocks. The embedded reconfigurable blocks can be either coarse grained (word level granularity) or FPGA like (bit level granularity). The different processing elements may communicate

87

N.S. Voros and K. Masselos (eds.), System Level Design of Reconfigurable Systems-on-Chips, 87-105.
© 2005 Springer. Printed in the Netherlands.

through a bus, however current trends are more towards communication networks on chip (for scalability, flexibility and power consumption issues).

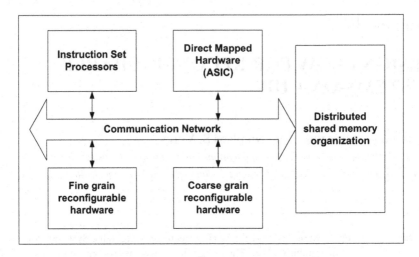

Figure 4-1. Abstract view of targeted implementation platform

The design of a SoC with reconfigurable hardware is not a trivial task. To obtain an efficient implementation an extended design flow is needed in order to cope with the reconfiguration aspects on a wide scale of commercially available platforms. In addition, a high abstraction level methodology needs to be developed for helping in deciding the instances of the implementation technologies, both for fine grained and coarse grained reconfigurable hardware. The requirements and the principles of such design methodology are further discussed in the rest of this chapter. It must be noted that the design flow and high level design methods described in the rest of this chapter can be equally apply to off-the-shelf system level FPGAs that include embedded hardwired blocks (including software processors and ASIC blocks).

2.   DESIGN FLOW REQUIREMENTS FOR RECONFIGURABLE SYSTEMS-ON-CHIP

The introduction of reconfigurable resources in Systems-on-Chip creates the need for modifications and extensions to conventional design flows with emphasis on the higher abstraction levels, where most important design decisions are made. In this section, conventional system level design flows are briefly presented and then system level design flow requirements for reconfigurable Systems-on-Chip are discussed.

2.1 Overview of conventional system level design flows

Driven by the SoC design growth, the demand for system level co-design methodologies is also increasing [6]. Academic and commercial sources have provided co-design methodologies/tools for a variety of application domains, with many hardware/software partitioning opportunities, synthesis, simulation and validation mechanisms, at different degrees of automation and levels of maturity.

As far as system specification is concerned, a variety of languages (HDL, object oriented, proprietary) are being used for system level specification. Some methodologies exploit a combination of languages in order to properly describe the hardware or software parts of the design. The trend is however to unify the system design specification in one description language capable of representing the system at the high level of abstraction [6].

The goal of hardware/software partitioning is the optimized distribution of system functions among software and hardware components. With respect to that, most beneficial are the methodologies that provide the partitioning at different levels of modeling without the necessity of rewriting the hardware or software specifications. This not only reduces the design iteration steps, but also enables easy inclusion of predefined library elements or IP blocks.

The important feature that should be taken into account during co-synthesis is the possibility of interface synthesis. The different possible inter-process communication primitives are covered in different methodologies. They are either fixed to the particular methodology or with the optional possibility of creating new primitives based on the existing ones.

Co-simulation techniques range from commercial simulation based on methodology specific simulation engines to combination of multiple simulation engines. Most of the methodology dependent co-simulators are based on event driven simulation, while some of them come with an option for co-simulation with other simulators [8].

Co-verification is mainly simulation based, meaning that the results of the HDL, ISS or proprietary simulations at different levels of the co-design flow are compared for correct functionality and timing, with the initial specifications. Debugging is enabled in some methodologies by exploiting a graphical tool or a proprietary user interface environment.

The main features of representative system level hardware/software co-design methodologies are summarized in Table 4-1.

As a natural consequence of what has been mentioned in the previous paragraphs, it is concluded that a generic traditional system level design flow usually involves the following key phases:

- System specification
- Hardware/software partitioning and mapping
- Architecture design
- System level (usually bus cycle accurate) simulation and
- Fabrication of hardware and software using tools provided by technology vendors.

Table 4-1. Summary of the main features of system level hardware/software co-design methodologies

	System Specification	HW/SW Partitioning	Co-synthesis	Co-simulation Co-verification	Remark
OCAPI-XL	Using C++ for functionality and architectural properties	Concurrent processes, partitioning on these processes can be made anywhere in the design flow	Interface synthesis and industrial tools for RTL synthesis	Unified co-simulation environment, performance estimation, co-simulation with other simulation engines	Future versions build on top of SystemC
SystemC	Using SystemC (based on C++) for functionality and architecture	System specifications can be refined to mixed SW and HW implementations	Channels, interfaces and events enable to model communication and synchronization	Simulation engine included, performance estimation	Becoming industry standard
VCC	Using C/VHDL for architecture and functionality	Template models of architecture where software and hardware area mapped		Unified co-simulation environment with emphasis on performance estimation	Perhaps the most complete tool set
Chess/ Checkers	Using proprietary nML language for processor architecture, C for application			Retargetable instruction set simulator simulates the execution of code on target processor	Useful for the design of embedded processors

continued

	System Specification	HW/SW Partitioning	Co-synthesis	Co-simulation Co-verification	Remark
CHINOOK	Using Verilog for functionality and pre-defined components for architecture	Allocate functionality to processors	Interface synthesis and industrial tools for RTL synthesis	HW/SW co-simulation engine included	Includes interface synthesis but requires tool specific models of processors and buses
COOL	Using a subset of VHDL for architecture and functionality	Synthesis and compilation tools used to compute the value for the cost metrics; specific algorithms to solve the HW/SW partitioning	Netlist & controllers for communication between HW and SW generated in VHDL	Commercial VHDL simulator to simulate functionality of the system specification and its implementation after co-synthesis	Precise modeling of cost and performance metrics
COSYMA	One processor and Verilog functionality	Allocate all to SW, then move slowest parts to HW.	Interface synthesis, and included tools for RTL synthesis	HW/SW co-simulation engine included	Applicable only to one processor architecture with hardware co-processor
N2C	C/C++, SystemC for system level description, extended C for hardware.	Manual	Automatic interface synthesis and industrial tools for RTL synthesis	Simulation environment; co-simulation with commercially available instruction set simulators	Support for IP cores
Esterel	Esterel language			BDD and temporal logic based verification techniques	Compilation of Esterel programs into FSM, HW or C programs

continued

	System Specification	HW/SW Partitioning	Co-synthesis	Co-simulation Co-verification	Remark
LYCOS	Subset of C for SW, subset of VHDL for HW	Different partitioning models and algorithms available	HW/SW communication through memory mapped I/O		Experimental co-synthesis environment
MESH	Textual		Modeling three independent layers for SW, scheduler/ protocol and HW resource		Project in early research phase
Ptolemy	Many models of computations that can be used in single design		Some code-generation tools	Powerful co-simulation engine for different models of computation	Features vary with models of computation

2.2   System level design flow requirements for reconfigurable Systems-on-Chip

The way in which the presence of embedded reconfigurable resources affects the major stages of a system level design flow, and the additional requirements it creates are discussed in this subsection.

2.2.1 System specification

In the system specification phase, the requirements, restrictions and specifications are gathered as when not using reconfigurable resources, but extra effort must be spent on identifying parts of the applications that serve as candidates for implementation with reconfigurable hardware. The incorporation of reconfigurable hardware brings new aspects to the architecture design task and to the partitioning and mapping task. In the architecture design task, a new type of architectural element is introduced. In architectural design space, the reconfigurable hardware can be viewed as being a time slice scheduled application specific hardware block. One way of incorporating reconfigurable parts into an architecture is to replace some hardware accelerators with a single reconfigurable block. The effects of reconfigurable blocks on the area, speed and power consumption should be completely understood before they can be efficiently used.

2.2.2 Hardware/software partitioning and mapping

During this phase, a new dimension is added to the problem. The parts of the targeted system that will be realized on reconfigurable hardware must be identified. There are some rules of thumb that can be followed to give a simple solution to this problem:

- If the application has several roughly same sized hardware accelerators that are not used in the same time or at their full capacity, a dynamically reconfigurable block may be a more optimized solution than a hardwired logic block.
- If the application has some parts in which specification changes are foreseeable, the implementation choice may be reconfigurable hardware.
- If there are foreseeable plans for new generations of application, the parts that will change should be implemented with reconfigurable hardware.

Furthermore, for the design of reconfigurable hardware instead of considering just area, speed and power consumption – as it happens in traditional hardware design – *the temporal allocation* and *scheduling problem* must also be addressed. This is achieved in a way similar to the policies followed for software tasks running on a single processor. This leads to increased complexity in the design flow, since the cost functions of the functionality implemented with reconfigurable technology include the problems of both hardware and software design.

There are basically two partitioning/mapping approaches supported by the existing commercial design flows: (a) the *tool oriented design flow*, and (b) the *language oriented design flow*. Examples of tool oriented design flows are the N2C by CoWare [7] and VCC by Cadence [5]. The design flows supported by these tools work well on traditional hardware/software solutions. Nevertheless, the refinement process of a design from unified and un-timed model towards RTL is tool specific, and the incorporation of new reconfigurable parts is not possible without unconventional trickery. Examples of language oriented design flows are OCAPI-XL [12] and SystemC [13]. Especially for the latter, since it promotes the openness of the language and the standard, the addition of a new domain can be made to the core language itself. However, the method mostly preferred is to model the basic constructs required for modeling and simulation of reconfigurable hardware, using basic constructs of the language. In this way, the language compatibility with existing tools and designs is preserved. SystemC extensions for reconfigurable hardware design and OCAPI-XL are thoroughly covered in Chapters 5 and 6 respectively.

2.2.3 Architecture design

A design flow that supports system descriptions at high abstraction level, must also support the reconfigurable technologies of different types and vendors. The main question that must be answered, even at the highest level of abstraction, is: *What to implement with reconfigurable technology and which reconfigurable technology to use?*[2] The design flow may answer these questions by using different techniques. First, analysis based tools compile the unified representation of the application functionality and produce information on which parts of the application are never run in parallel. This information can be used to determine what functionality can be implemented in different contexts of a reconfigurable block.

An alternative method is the use of cost functions for each implementation technology. Cost functions help in making quick design decisions using several parameters and optimization criteria at the same time. Another category of tools use profiling information gathered in simulations in order to partition the application and to produce a context scheduler to be used in the final implementation. Example of this approach is a toolset for MorphoSys [14] reconfigurable architecture.

Finally, the most realistic alternative for industrial applications is the simulation based approach. In this approach, the partitioning, mapping and scheduling are accomplished manually by the designer, while the results and the efficiency are verified through simulations. This approach is also the easiest to incorporate into an existing flow, since the required tool support is limited compared to the previous approaches. This also leaves all the design decisions to the designer, which is preferred by many industrially used design flows.

When considering designing additions to a language or a tool that can support modeling and simulation of reconfigurable technologies, a set of parameters that differentiate the implementation technologies need to be identified: (a) the reconfigurable block capacity in gates, (b) the amount of context memory required to hold configurations, (c) the reconfiguration time and support for partial reconfiguration, (d) typical clock or transaction speed, and (e) power consumption information. The

---

[2] A brief introduction to existing reconfigurable hardware technologies is presented in Chapter 3.

aforementioned parameters are adequate for modeling any type of homogenous reconfigurable technology. The simulation accuracy resulting from using these parameters is not optimal, but it is sufficient for giving the designer an idea of how each different reconfigurable technology affects the total system performance.

The results needed for steering the design space exploration, and verifying that the design decisions fulfill the total system performance, are:

- *Spatial utilization*, which is needed to validate the correct size of the block and also granularity of the contexts.
- *Temporal utilization*, that is measured to compare the time spent in configuring the block, waiting for activation and actively doing the computation.
- *Context memory bus load*, which is measured to analyze the effects of the reconfiguration memory bus traffic on the performance of system buses.
- *Area* and *power consumption* which are compared against hardware or software implementation.

The aforementioned results should be used as additional information in order to decide which reconfigurable technology to use and which parts of the application will be implemented with it.

When comparing the requirements pertaining to reconfigurability in existing design flows, it can be seen that the existing design flows and tools do not support any of the requirements directly. Either the tools and languages should be improved or company specific modifications are needed [1, 2, 3].

3.   THE PROPOSED DESIGN FLOW FOR RECONFIGURABLE SoCs

This section provides the general framework of the proposed design flow for designing complex SoCs that contain reconfigurable parts. The flow aims to improve the design process of a SoC in order to use the available tools in an optimal way [11].

The main idea of the design flow proposed is to identify the parts of a co-design methodology, where the inclusion of reconfigurable technologies has the greatest effect. This is very important since there are no commercial tools or methodologies to support reconfigurable technologies, yet. The design flow is divided in three parts as shown in Figure 4-2. The *System-Level Design (SLD)* refers to the high level part of the proposed flow, while the *Detailed Design (DD)* and *Implementation Design (ID)* refer to the back end part of the methodology.

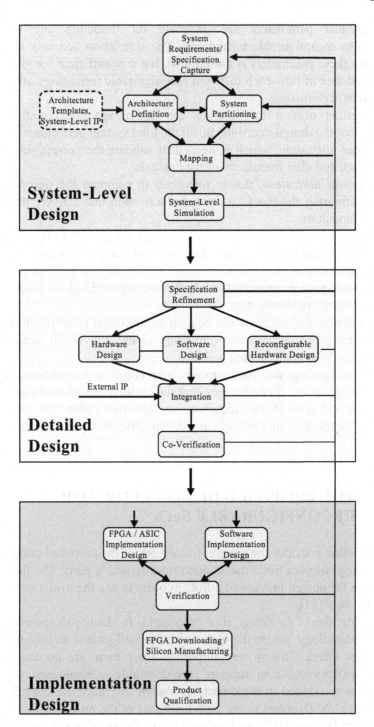

Figure 4-2. The proposed Design Flow

Details on the formalisms used are thoroughly covered in Chapters 5 and 6, while Chapters 7, 8 and 9 provide information how the proposed framework can be applied for the design of real world case studies.

3.1 System Level Design (SLD)

At the SLD phase, the main targets are:
- to develop a specification of the application associated with the requirements captured (and analyzed),
- to design the architecture of the SoC,
- to select major implementation technologies,
- to partition the application for implementation in hardware, software or reconfigurable hardware and,
- to evaluate the performance of the partitioned system.

The requirements are captured and analyzed in the specification phase and the results are fed to the next phases of the design flow. Architecture templates can be used to derive an initial architecture. They can be based on previous versions of the same product, a different product in the same product family, a design/implementation platform provided by the design tool or semiconductor vendor or even on information of a similar system by a competitor.

At the architecture definition phase, bus cycle accurate models of the architectural units are created, so that the performance of the architecture can be evaluated using system level simulations.

In the partitioning phase, the functional model of the application is partitioned in software, hardware and reconfigurable hardware. These partitions are then mapped onto the architecture, annotated with estimations of timing and other characteristics needed in the mapping phase.

At the SLD, the reconfiguration issues emerge in the following forms:
- The goals for reconfiguration (e.g. flexibility for specification changes and performance scalability) with associated constraints are identified at the requirements and specification step.
- At the design space exploration step, the reconfigurable hardware manifests itself as a computing resource in a similar way as an instruction set processor or a block of fixed hardware, thus bringing a new dimension to the design space exploration.

3.2 Detailed Design (DD)

At the DD phase, the specifications are refined and verification is planned according to targeted implementation technologies, processors etc.

The design tools used are fixed according to the selected processors and the chosen reconfigurable and fixed hardware technologies. Additionally, the verification and testing strategy are planned. After this, the individual partitions of hardware, software and reconfigurable hardware are designed and verified.

When all parts are finished, the designed modules of hardware, software and reconfigurable hardware are integrated into a single model. In the co-verification step, the functionality of the integrated model is checked against the reference implementation or the executable specification. Moreover, implementation related issues like timing and power consumption are modeled. If the results are satisfactory, the design is moved to the Implementation Design phase, otherwise iterations to Detailed Design or even to System Level Design phases are required.

At the DD, the reconfiguration issues emerge in the following ways:

- At the specification refinement and technology specific design, the reconfigurable hardware requires communication mechanisms to software and/or fixed hardware to be added; in case of dynamic reconfiguration mechanisms to handle context multiplexing are also needed.
- The integration and co-verification combines the reconfigurable hardware components with other hardware and software components onto a single platform that accommodates also external IP (e.g. processor, memory and I/O sub-system models) and provides co-verification of the overall design. The reconfigurable hardware is simulated in a HDL simulator or emulated in an FPGA emulator.
- Specific HDL modeling rules need to be followed for multiple dynamically reconfigurable contexts [2, 3].
- The reconfigurable hardware modules must be implemented using the selected technology, including the required control and support functions for reconfiguration.
- In the integration and verification phases, the vendor specific design and simulation/emulation tools must be used.

3.3 Implementation Design (ID)

At the ID, the reconfiguration issues emerge in the following forms:

- Dynamic reconfiguration requires configuration bit streams of multiple contexts to be managed.
- Specific design rules and constraints must be followed for multiple dynamically reconfigurable contexts [2, 3].

4. RECONFIGURATION ISSUES IN THE PROPOSED DESIGN FLOW

As indicated in the previous section, there are several issues regarding reconfiguration. The next sections emphasize how these aspects are addressed in the context of the proposed design framework. The focus is on system level design issues, although detailed and implementation design apsects are briefly discussed to complete the picture.

4.1   Reconfiguration issues at System Level Design

4.1.1 Needs and Requirements for Reconfiguration

The requirements and specification capture identifies the required functionality, performance, critical physical specifications (e.g. area, power) and the development time required for the system. All the aforementioned characteristics are described in the form of an executable model, where the goals for reconfiguration (e.g. flexibility for specification changes and performance scalability) are identified as well.

In general, simultaneous flexibility and performance requirements form the basic motivation for using reconfiguration in System-on-Chip designs. Otherwise either pure software or fixed hardware solutions could be more competitive. Reconfigurable technologies are a promising solution for adding flexibility, while not sacrificing performance and implementation efficiency. They combine the capability of post fabrication functionality modification with the spatial/parallel computation style.

The inclusion of reconfigurable hardware to a telecommunication system may introduce significant advantages both from market and implementation points of view:

- Upgradability
  - Need to conform to multiple or migrating international standards
  - Emerging improvements and enhancements to standards
  - Desire to add features and functionality to existing equipment
  - Service providers are not sure what types of data services will generate revenue in the wireless communications world
  - Introduction of bug fixing capability for hardware systems.
- Adaptivity
  - Changing channel, traffic and applications
  - Power saving modes.

Although the reconfigurable hardware is beneficial in many cases, significant overheads may also be introduced. These are mainly related to the time required for the reconfiguration and to the power consumed for reconfiguring a system. Area implications are also introduced (memories storing configurations, circuits required to control the reconfiguration procedure).

The requirements capture should identify and define the following reconfiguration aspects:
- Type of reconfiguration wanted in the system
  - Static or dynamic (single or multiple contexts)
  - Level of granularity (from coarse to fine)
  - Style of coupling (from loosely to closely coupled).
- Requirements and constraints on system properties (performance, power, cost, etc)
- Requirements and constraints on design methodology (pre-defined architecture, pre-selected technologies and IPs, tools, etc)

The information outlined above is needed in the later stages of the design flow. However, the techniques for identification of needs and capture of requirements are company specific.

4.1.2 Executable Specification

The specification capture is similar to the case of systems that employ only traditional hardware. The functionality of the system is described using a C-like formalism e.g. SystemC, OCAPI-XL. The executable specification can be used for several purposes:
- The test bench used in all phases of the design flow can be derived from the executable specification.
- The compiler tools and profiling information may be used to determine which parts of an application are most suitable for implementing with dynamically reconfigurable hardware. This is achieved in the partitioning phase of the design flow.
- The ability to implement executable specification validates that the design team has sufficient expertise on the application.

Executable specification is a must in order to be able to tackle reconfigurability issues at the system level design.

4.1.3 Design Space Exploration

The design space exploration phase analyses the functional blocks of the executable model with respect to reconfigurable hardware implementations. More specifically:
- It defines architecture models containing reconfigurable resources based on templates.

- It decides the system partitioning onto reconfigurable resources (in addition to hardware and software) based on the analysis results.
- It maps the partitioned model onto selected architecture models.
- It performs system level simulation to estimate the performance and resource usage of the resulting system.

The architecture of the device is defined partly in parallel and partly using the system specification as input. The initial architecture depends on many factors in addition to the requirements of the application. For examples a company may have experience and tools for certain processor core or semiconductor technology, which restricts the design space. Moreover, the design of many telecom products does not start from scratch, since they implement advanced versions of existing devices. Therefore the initial architecture and the hardware/software partitioning is often given at the beginning of the system level design. There are also cases where the reuse policy of each company mandates designers to reuse architectures and code modules developed in previous products. The old models of an architecture are called *architecture templates*.

As far as dynamic reconfiguration is concerned, it requires partitioning to address both temporal and spatial dimensions. Automatic partitioning is still an unsolved problem, but in specific cases solutions for temporal partitioning [4], task scheduling and context management [10] have been proposed. In the context of industrial SoC design, however, the system partitioning is mostly a manual effort. Based on the needs and requirements for reconfiguration, the executable specification is analyzed in order to identify parts that could gain benefits from implementation on reconfigurable resources. This analysis can be supported by estimations of performance and area done with respect to pre-selected technologies, architectures and IPs, e.g. specific ISP and reconfigurable technology.

During the mapping phase, the functionality defined in executable specification is refined according to the partitioning decisions so that it can be mapped onto the defined architecture. In order to include in the system level simulation the effects of the chosen implementation technology, different estimation techniques can be used:

- Software parts may be compiled for getting running time and memory usage estimates.
- Hardware parts may be synthesized at high level to get estimates of gate counts and running speed.
- The functional blocks implemented with reconfigurable hardware are also modelled so that the effects of reconfiguration can be estimated.

Finally simulations are run at the system level, to get information concerning the performance and resource usage of all architectural units of the device.

Efficient design space exploration is the core of the proposed design framework. With respect to the design of reconfigurable systems parts, it supports:

- Early estimation of function blocks/processes for performance (hardware, software and reconfigurable), cost (area) etc.
- System partitioning, especially multi context partitioning and scheduling
- Architecture definition
- Mapping
- Performance evaluation.

4.2   Reconfiguration issues at Detailed Design

The specification refinement and technology specific design transform the functional blocks of the executable model to design components targeting reconfigurable hardware (in addition to hardware and software) according to the partitioning decisions. Important issues at this stage include iterative improvements in hardware, software and reconfigurable hardware specification. The designers take into account not only design (modeling language, targeted platform, co-simulation and testing strategy), but also economical and product support aspects of the design, exploiting the specific reconfigurable hardware features.

The integration phase combines the hardware, software and reconfigurable hardware components into a single platform that accommodates also external IP e.g. processor, memory, I/O sub-system models. The integration phase considers two different approaches: *language based approach* (SystemC, OCAPI-XL) and *tools oriented approach* (CoWare N2C) to combine the heterogeneous components of the target system on a single platform.

The reconfigurable hardware requires communication mechanisms to software and/or fixed hardware to be added. Different types of mechanisms can be chosen to handle communication between the components: memory based communication, bus based, coprocessor style and even datapath integrated reconfigurable functional units. Bus based communication between the components requires specific interfaces for both the reconfigurable fabric and hardware/software sides of the system. On the software side, drivers are required to turn software operations into signals on the hardware. On the FPGA fabric and hardware side, interfaces to the system bus must be built. The FPGA fabric and CPU can also communicate directly by shared memory.

Regarding the software and fixed hardware design flows, they do not differ from traditional ones. For statically reconfigurable hardware the design flow is similar to that of fixed hardware. For dynamically reconfigurable hardware, the module interfaces, communication and synchronization are designed according to the principles of a context scheduler. Specific HDL modeling rules need to be followed for multiple dynamically reconfigurable contexts [3, 9]. In the case of dynamic reconfiguration, mechanisms to handle context multiplexing are also needed. A high level scheme for describing dynamic reconfiguration should address how dynamically reconfigurable circuits compose with other circuits over a bus structure.

4.3 Reconfiguration issues at Implementation Design

Reconfiguration partitions the application temporally and multiplexes in time the programmable logic to meet the hardware resource constraints. When reconfiguration takes place at run time, the reconfiguration time is part of the run time overhead and has to be minimized. Also, multiple reconfiguration bit streams need to be stored for the different contexts being multiplexed onto the programmable logic. This problem is exacerbated for System-on-Chip implementations where the entire application needs to be stored in on-chip memory.

When multiple context reconfigurable techniques are considered [3, 9], dedicated partitioning and mapping techniques are applied during System Level Design phase. Later, during Implementation Design step, an inter-context communication scheme has to be provided. Inter-context communication refers to how data or control information is transferred among different contexts. Usually, transfer registers are used for interconnecting between the previous – last, and current – next context. Backup registers are also used to store the status values when the context switches out and later switches in. When bulk buffers are more practical for inter-context communication, memory regions can be allocated anywhere in the chip by using memory mode of the reconfigurable cells. These memory regions can be accessed from all the contexts as shared buffers. It is instructive to compare this high bandwidth for inter-context communication with a multiple FPGA situation, where bandwidth is inherently limited to external pins. The huge bandwidth makes multi-context partitioning much easier than the multi-FPGA partitioning.

5. CONCLUSIONS

The design flow for reconfigurable SoCs presented in the previous sections is divided in three phases: In the *system level design phase*, where the requirements and specifications are captured; functionality in the form of executable specification is analyzed, partitioned and mapped onto the architecture, and the performance of the system is validated. In the *detailed design phase*, the communication and modules are refined and transformed, integrated and co-verified through co-simulation or co-emulation. The implementation design maps the design onto the selected implementation platform. The implementation technologies treated in this methodology are software executed in an instruction set processor, traditional fixed hardware and dynamically reconfigurable hardware.

Emphasis is given on the system level part of the design flow where methods for the modeling and simulation of reconfigurable hardware parts of a reconfigurable SoC are required. Methods and tools towards this direction are presented in Chapters 5 and 6 respectively.

REFERENCES

1. ADRIATIC Project IST-2000-30049 (2002) Deliverable D2.2: Definition of ADRIATIC High-Level Hardware/Software Co-Design Methodology for Reconfigurable SoCs. Available at: http://www.imec.be/adriatic
2. ADRIATIC Project IST-2000-30049 (2003) Deliverable D3.2: ADRIATIC back-end design tools for the reconfigurable logic blocks. Available at: http://www.imec.be/adriatic
3. ADRIATIC Project IST-2000-30049 (2004) Addendum to Deliverable D3.2: ADRIATIC back-end design tools for the reconfigurable logic blocks. Available at: http://www.imec.be/adriatic
4. Bobda C (2003) Synthesis of Dataflow Graphs for Reconfigurable Systems using Temporal Partitioning and Temporal Placement. PhD Dissertation, University of Paderborn
5. Cadence (2004) http://www.cadence.com/datasheets/vcc_environment.html
6. Cavalloro P, Gendarme C, Kronlof K, Mermet J, Van Sas J, Tiensyrja K, Voros NS (2003) System Level Design Model with Reuse of System IP, Kluwer Academic Publishers
7. CoWare Inc (2004) Available at: http://www.coware.com
8. Gioulekas F, Birbas M, Voros NS, Kouklaras G, Birbas A (2005) Heterogeneous System Level Co-Simulation for the Design of Telecommunication Systems. Journal of Systems Architecture (to appear), Elsevier
9. Keating M, Bricaud P (1999) Reuse Methodology Manual. Second Edition, Kluwer Academic Publishers
10. Maestre R, Kurdahi FJ, Fernandez M, Hermida R, Bagherzadeh N, Singh H (2001) A framework for reconfigurable computing: task scheduling and context management. IEEE Transactions on Very Large Scale Integration (VLSI) Systems, vol. 9, issue 6, pp. 858 – 873
11. Masselos K, Pelkonen A, Cupak M, Blionas, S (2003) Realization of wireless multimedia communication systems on reconfigurable platforms. Journal of systems architecture, vol. 49 (2003) no: 4–6, pp. 155–175

12. OCAPI-XL (2004) Available at: http://www.imec.be/ocapi/welcome.html
13. SystemC (2004) Available at: http://www.systemc.org
14. Tiwari V, Malik S, Wolfe A, Lee MTC (1996) Instruction level power analysis and optimization of software. Journal of VLSI Signal Processing, Kluwer Academic Publishers, pp. 223 – 238

Chapter 5

SYSTEMC BASED APPROACH

Yang Qu and Kari Tiensyrjä
VTT Electronics, P.O.Box 1100, FIN-90571 Oulu, Finland

Abstract: This chapter describes the SystemC based modelling techniques and tools that support the design of reconfigurable systems-on-chip (SoC). For designing of reconfigurable parts at system level, we developed: 1) an estimation method and tool for estimating the execution time and the resource consumption of function blocks on dynamically reconfigurable logic to support system partitioning, 2) a SystemC based modeling method and tool for reconfigurable parts to allow fast design space exploration through 3) system-level simulation using transaction-level models of the system.

Key words: Configuration overhead; context switching; design space exploration; dynamic reconfiguration; estimation; mapping; partitioning; reconfigurable; reconfigurability; SystemC; system-on-chip; workload model.

1.   INTRODUCTION

Reconfigurability does not appear as an isolated phenomenon, but as a tightly connected part of the overall SoC design flow. The SystemC-based approach is therefore not intended to be a universal solution to support the design of any type of reconfigurabily. Instead, we focus on a case, where the reconfigurable components are mainly used as co-processors in SoCs.

SystemC 2.0 is selected as the backbone of the approach since it is a standard language that provides designers with basic mechanisms like channels, interfaces and events to model the wide range of communication and synchronization found in system designs. More sophisticated mechanisms for the system-level design can be built on top of the basic constructs. Due to the standard language and open source reference implementation, SystemC 2.0 has become a language of choice for a growing number of system architects and system designers.

N.S. Voros and K. Masselos (eds.), System Level Design of Reconfigurable Systems-on-Chips, 107-131.
© 2005 *Springer. Printed in the Netherlands.*

The SystemC based approach covers the reconfiguration extension and the related methods and tools that can be easily embedded into a SoC design flow. The system-level design part of the design flow presented in Chapter 4 is shown in Figure 5-1.

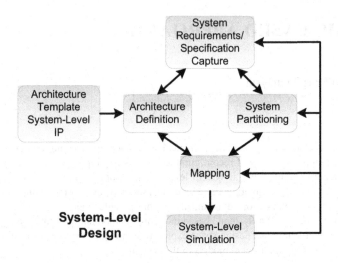

Figure 5-1. System-level design part of proposed design flow.

The following new features are identified in each phase of system-level design when reconfigurability is taken into account:

- *System Requirements and Specification Capture* needs to identify requirements and goals of reconfigurability.
- *Architecture Definition* needs to treat the reconfigurable resources as abstract models and include them in the architecture models.
- *System Partitioning* needs to analyze and estimate the functions of the application for software, fixed hardware and reconfigurable hardware.
- *Mapping* needs to map functions allocated to reconfigurable hardware onto the respective architecture model.
- *System-Level Simulation* needs to observe the performance impacts of architecture and reconfigurable resources.

In the SystemC based approach, we assume that the design does not start from scratch, but it is a more advanced version of an existing device. The new architecture is defined partly based on the existing architecture and partly using the system specification as input. The initial architecture is often dependent on many things not directly resulting from the requirements of the application. The company may have experience and tools for certain processor core or semiconductor technology, which restricts the design space and may produce an initial hardware/software (HW/SW) partition.

Therefore, the initial architecture and the HW/SW partition are often given in the beginning of the system-level design. The SystemC extension is designed to work with a SystemC model of the existing device to suit the design considering dynamically reconfigurable hardware Figure 5-2 (a) gives a graphical view of the initial architecture, and Figure 5-2 (b) shows the modified architecture with using the SystemC based extensions.

The way that the SystemC based approach incorporates dynamically reconfigurable parts into architecture is to replace SystemC models of some hardware accelerators with a single SystemC model of reconfigurable block. The objective of the SystemC based extensions is to provide a mechanism that allows designers to easily test the effects of implementing some components in the dynamically reconfigurable hardware. The provided supports in the SystemC based approach include:

- Analysis support for design space exploration and system partitioning.
- Reconfigurability modelling by using standard mechanisms of SystemC.
- System-level simulation using transaction-level models of the application workload and the architecture.

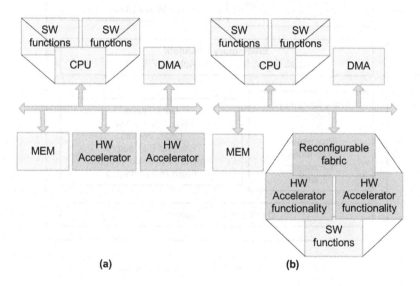

(a)   (b)

Figure 5-2. (a) Typical SoC architecture and (b) modified architecture using dynamically reconfigurable hardware.

2.   SYSTEMC 2.0 OVERVIEW

SystemC is a standard modelling language based on C++. Its version 1 provides a class library that implements objects like processes, modules, ports, signals and data types for hardware modelling. The model is compiled by a standard C++ compiler for execution on an event based simulation kernel. The version 2 introduces a language architecture shown in Figure 5-3 [1]. It provides core language constructs like channels, interfaces and events for system-level modelling. Elementary and more sophisticated channels can be built using the core language to support various communication, synchronization and model of computation paradigms.

The basic system-level constructs of the language are introduced in following sections, but for more complete information it is advisable to read the Functional Specification for SystemC 2.0 [2].

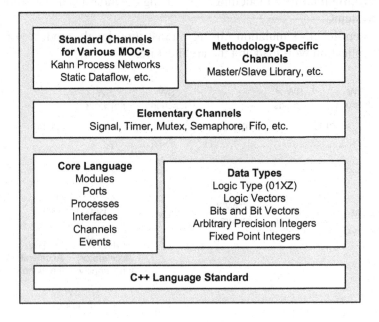

Figure 5-3. SystemC language architecture.

2.1   Channels

SystemC 2.0 channels implement one or many interfaces and they contain the functionality of the communication. Channels are used especially in designing and simulating functionality of buses. Functionality such as addresses, addressing schemes, priorities buffer sizes etc. can be configured runtime and therefore the effect of these design decisions can be simulated easily without large modifications to the code.

Also, since it is possible to attach multiple ports to an interface the number of bus masters or slaves can be chosen in compile time without modifying the bus code. When system level modules are implemented correctly for use of parameters and variable number of connected ports, design space exploration becomes an easy task.

2.2   Ports and Interfaces

The model of communication in SystemC 2.0 can be more abstract than in register-transfer level (RTL) description. User can define a set of interface methods that modules use for communication. For example a system level model of a memory controller can contain three interface methods, a read method, a write method and a burst read method. The actual behavioural implementation of a method is left to the module that provides the interface. The module that uses an interface does this via a port. This way the detailed implementation of an interface can be separated from the object that is using the interface. Using interfaces makes it also simpler to simulate and measure the effect of for example burst reading to the performance of a system. This is called transaction level modelling (TLM).

2.3   Events and Dynamic Sensitivity

Events are low-level synchronization mechanisms. They can be used to transfer control from one process to another. The effect can occur immediately, after next delta cycle or after some defined time. Dynamic sensitivity in SystemC 2.0 means that a process can alter its sensitivity list during runtime. Process can wait any set of events or time making for example design and simulation of state machines easy and errors are reduced since the sensitivity list can be suppressed in each state to minimum.

3.   OVERVIEW OF SYSTEMC BASED EXTENSIONS

Since SystemC promotes the openness of the language and the standard, the addition of new domain can be made to the core language itself. However, a preferred method is to model the basic constructs required for modelling and simulation of reconfigurable hardware (RHW) using basic constructs of the language and therefore preserving the compatibility with existing tools and designs. For this reason, the extension does not intend to extend the System 2.0 language itself.

The terms and concepts specific to the SystemC based approach used in the following sections are defined as follows:

- *Candidate Component*: Candidate components denote those application functions that are considered to gain benefits from their implementation on a reconfigurable hardware resource. The decision whether a task should be a candidate component is clearly application dependent. The criterion is that the task should have two features in combination: flexibility (that would exclude an ASIC implementation) and high computational complexity (that would exclude a software implementation). Flexibility may come either from the point that the task will be upgraded in the future or in view of hardware resources sharing with other tasks with non-overlapping lifetimes for global area optimization.

- *Dynamically reconfigurable fabric (DRCF)*: The dynamically reconfigurable fabric is a system-level concept that represents a set of candidate components and the required reconfiguration support functions, which later on in the design process can be implemented on a reconfigurable hardware resource.

- *DRCF component*: The DRCF component is a transaction-level SystemC module of the DRCF. It consists of functions, which mimic the reconfiguration process, and the instances of SystemC modules of the candidate components to present their functionality during system-level simulation. It can automatically detect reconfiguration request and trigger the reconfiguration process when necessary.

- *DRCF template*: The DRCF template is an incomplete SystemC module, from which to create the DRCF component.

The SystemC based extensions [3] are highlighted in the modified version of the System-Level Design diagram as shown in Figure 5-4. The three focuses are estimation support, DRCF modelling method and system simulation.

- The estimation approach [4] is based on a prototype tool that can produce the estimates of software execution time on an instruction-set processor (ISP) and the estimates of hardware execution time and resource consumption on an FPGA. The estimates provide information for system partitioning and selection of candidate components. When a full SW/HW/RHW system partitioning is considered, traditional analysis methods and tools are still required.

- The DRCF modelling method [5, 6] focuses on the modelling of the reconfiguration overhead. Modelling the functionality of the candidate components that are mapped onto the reconfigurable resources is not affected by the extension. Different features associated with reconfiguration technology are not directly modelled. Instead, the model describes the behaviour of the reconfiguration process and relates the performance impact of the reconfiguration process to a set of parameters that are extracted and annotated from the reconfiguration technology. Thus, by tuning the parameters, designers can easily evaluate trade-offs among different technology alternatives and perform fast design space exploration at the system level.

- The system-level simulation is based on the transaction-level SystemC model and uses abstract workload and capacity models of application and architecture for performance evaluation and studying of alternative architectures and mappings.

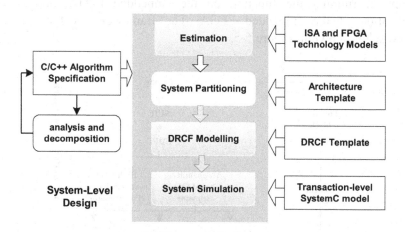

Figure 5-4. SystemC reconfigurability extensions for system-level design.

4. ESTIMATION APPROACH TO SUPPORT SYSTEM ANALYSIS

System analysis is applied in two phases in the SystemC based approach. In the first phase, it focuses on HW/SW partitioning and helps designers to create the initial architecture based on an agreed partitioning decision. The initial architecture sets the starting point from which the SystemC based approach produces the system-level model for the architecture including the DRCF component that is a corresponding SystemC model of the dynamically reconfigurable hardware with the modules to be implemented in it. In the second phase, system analysis focuses on studying the trade-off of performance and flexibility and helps designers to identify candidate components to be implemented in the dynamically reconfigurable hardware. System analysis is performed by designers mainly based on their experience, which may not produce reliable results in all cases especially if designers have to carry out system analysis from the scratch. In this section, an estimation approach to support the work of system analysis is presented.

The estimation approach focuses on a reconfigurable architecture in which there is a RISC processor, an embedded FPGA, and a system bus as a communication channel. It starts from function blocks represented using C-language and produces the following estimates for each function block: software execution time in terms of running the function on the RISC core, mappability of the function and the RISC core, hardware execution time in terms of running the function on the embedded FPGA, and resource utilization of the embedded FPGA. The framework of the estimation approach is shown in Figure 5-5.

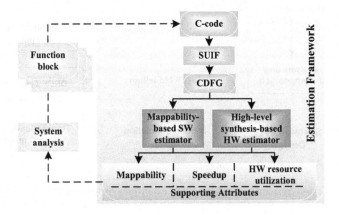

Figure 5-5. Estimation framework.

Blocks inside the shaded area are the functions performed by the estimation approach, and data representations used by the estimation approach. Detailed explanations are given in the following sections. Outside the shaded area, the blocks with the name "Function block" serves as input to the estimation approach. These function blocks can either be the results from system decomposition, with the granularity decided by designers, or they can be the corresponding SystemC modules from the initial architecture. In the former case, the estimation approach is meant for the first phase of system analysis, which is to help designers to make trade-off between hardware implementation and software implementation. In the latter case, the estimation approach is meant for the second phase of system analysis, which is to help designers to evaluate the trade-off between performance and flexibility when comparing fixed hardware implementation and dynamically reconfigurable hardware implementation.

Estimates of hardware resource utilization of the modules are fed into the SystemC extension as separate parameters.

4.1 Creation of Control/Data Flow Graph from C Code

Control/data flow graph (CDFG) is a combined representation of data flow graph (DFG), which exposes the data dependence of algorithms, and control flow graph (CFG), which captures the control relation of DFGs. C-based function block is used as the starting point and CDFG is used as the intermediate representation of the estimation approach. SUIF compiler [7] is used as a front-end tool to analyze the C code, and a purpose-specific code converter is used to transform the SUIF intermediate representation into CDFG. The main process in conversion is to find basic blocks, which contain only sequential executions without any jump in between, and to map each of them onto a single DFG and the jump statements between the basic blocks onto the control relation of DFGs. The characteristics of the C functions are studied though profiling, and the profiling data are attributes in the target CDFG.

4.2 High-Level Synthesis-Based Hardware Estimation

A graphical view of the hardware estimation is shown in Figure 5-6. Taking the CDFG with corresponding profiling information and a model of embedded FPGA as inputs, the hardware estimator carries out a high-level synthesis-based approach to produce the estimates.

Main tasks performed in the hardware estimator as well as in a real high-level synthesis tool are scheduling and allocation. Scheduling is the process in which each operator is scheduled in a certain control step, which is usually a single clock cycle, or crossing several control steps if it is a multi-cycle operator. Allocation is the process in which each representative in the CDFG is mapped to a physical unit, e.g. variables to registers, and the interconnection of physical units is established.

The embedded FPGA is viewed as a co-processing unit, which can independently perform a large amount of computation without constant supervision of the RISC processor. The basic construction units of the embedded FPGA are static random access memory (SRAM)-based look-up tables (LUT) and certain types of specialized function units, e.g. custom-designed multiplier. Routing resources and their capacity are not taken into account. The model of the embedded FPGA is in a form of mapping-table. The index of the table is the type of the function unit, e.g. adder. The value mapped to each index is hardware resources in terms of the number of LUTs and the number of specialized units, required for this type of function unit.

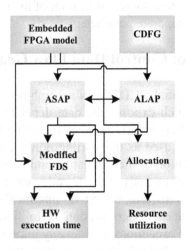

Figure 5-6. High-level synthesis-based hardware estimation.

As-soon-as-possible (ASAP) scheduling and as-late-as-possible (ALAP) scheduling [8] determine the critical paths of the DFGs, which together with the control relation of the CFGs are used to produce the estimate of hardware execution time. For each operator, the ASAP and ALAP scheduling processes also set the range of clock cycles within which it could be legally scheduled without delaying the critical path. These results are required in the next scheduling process, a modified version of force-directed-scheduling (FDS) [9], which intends to reduce the number of function units, registers and buses required by balancing the concurrency of the operations assigned to them without lengthening the total execution time. The modified FDS is used to estimate the hardware resources required for function units.

Finally, allocation is used to estimate the hardware resources required for interconnection of function units. The work of allocation is divided into 3 parts: register allocation, operation assignment and interconnection binding. In register allocation, each variable is assigned to a certain register. In operation assignment, each operator is assigned to a certain function unit. Both are solved using the weighted-bipartite algorithm, and the common objective is that each assignment should introduce the least number of interconnection units that will be determined in the last phase, the interconnection binding. In this approach, multiplexer is the only type of interconnection unit, which ease the work of interconnection binding. The number and type of multiplexers can be easily determined by simply counting the number of different inputs to each register and each function unit.

4.3 Mappability Based Software Estimation

Software estimator produces two estimates: software execution time, and mappability of an architecture-algorithm pair. A profile-directed operation-counting based static technique is used to estimate software execution time. The architecture of the target processor core is not taken into account in the timing analysis. The main idea of estimating the software execution time is as following. Firstly, the number of operations with each type is counted from the CDFG. Then, each type of operation nodes in the CDFG is mapped to one or a set of instructions of the target processor in a pre-defined manner. Then the total number of instructions is calculated from the results of the first two steps simply using multiplication and addition. Finally, with the assumption that these instructions are performed with an ideal pipeline, the software execution time is the multiplication result of the total number of instructions and the period of the clock cycle.

Mappability of an architecture-algorithm pair means the degree of matching between resources provided by the processor architecture and the requirements described by the algorithm [10]. The mappability estimate is calculated via a set of correlation functions, which take into account the instruction set, register structure, bus efficiency, branch effect, pipeline efficiency and parallelism. CAMALA is a prototype tool to study mappability of an architecture-algorithm pair. It takes CDFG as input and produces estimate of mappability within the range from 0 to 1. An optimal mapping is an exact mapping with a value of one, and both over-required resources and under-utilized resources are reflected as poor mapping results with values near zero.

4.4 Candidate Component Selection

Candidate component selection is an application-dependent procedure. When global resource saving is an issue, the resource estimates are important inputs. However, to make justified decisions, other information, such as power consumption should be included as inputs. More importantly, control/data dependence between candidate components should be analyzed. Obviously, there should be control dependence between candidate components that are mapped to different contexts. Current approach does not include automated tools to support the analysis. Other tools and manual analysis are the solutions for now.

5.   MODELLING RECONFIGURATION OVERHEAD

The modelling method of the DRCF focuses on how to represent the reconfiguration overhead and how to reveal its performance impact during system simulation.

The candidate components that are mapped onto the reconfigurable resources are hardware accelerator tasks. Reconfiguration is required when a called task is not loaded in the reconfigurable resources. The difference of handling incoming messages between tasks mapped to a fixed accelerator and tasks mapped to reconfigurable resources is shown in Figure 5-7.

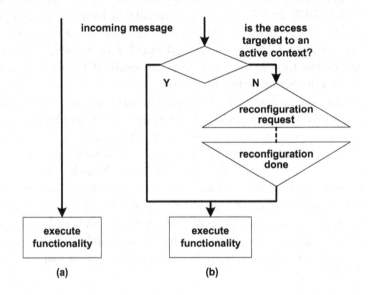

Figure 5-7. (a) Handling incoming messages as a fixed hardware accelerator (b) Handling incoming messages as a reconfigurable task.

The idea of the DRCF is to automatically capture the reconfiguration request and trigger the reconfiguration. In addition, a tool to automate the process that replaces candidate components by a DRCF component is developed, so system designers can easily perform the test-and-try and the design space exploration process is easier. In order to let the DRCF component be able to capture and understand incoming messages, the SystemC modules of the candidate components must implement the *read()*, *write()*, *get_low_addr()* and *get_high_addr()* interface methods showed in the code below.

```
class bus_slv_if: public virtual sc_interface
{
  public:
      virtual sc_uint<ADDW> get_low_addr() =0;
      virtual sc_uint<ADDW> get_high_addr() =0;
      virtual bool read(...) =0;
      virtual bool write(...) =0;
};
```

The DRCF component implements the same interface methods and conditionally calls the interface methods of target modules. In fact, these interface methods are very common for bus slave modules in transaction-level models.

5.1   Parameterized DRCF Template

The performance impact of using the dynamically reconfigurable hardware is dependent on the underlying reconfigurable technology. Products from different companies or different product families from the same company have very different characteristics, e.g. size of reconfigurable logic and granularity of reconfigurable logic.

Different features associated with the reconfigurable technology are not directly modelled in the DRCF component. Instead, the DRCF component contains the functions that describe the behaviour of the reconfiguration process and relates the performance impact of the reconfiguration process to a set of parameters. Thus, by tuning the parameters, designers can easily evaluate the trade-offs between different technologies without going into implementation details.

In the SystemC extension, a parameterized DRCF template is used. At the moment, the following parameters are available for designers:

- The memory address, where the context is allocated in the extra DRCF memory.
- The length of the required memory space, which represents the size of the context.
- Delays associated with the reconfiguration process in addition to delays of memory transfers.

5.2   DRCF Component and RSoC Model

A general model of a reconfigurable system-on-chip (RSoC) is shown in Figure 5-8. The left hand side depicts the architecture of the RSoC. The right hand side shows the internal structure of the DRCF component.

The DRCF component is a single hierarchical SystemC module, which implements the same bus interfaces in the same way as other HW/SW modules. A configuration memory is modelled, which could be an on-chip or off-chip memory that holds the configuration data. Each candidate component (F1 to Fn) is an individual SystemC module, which implements the top-level bus interfaces with separate system address space, and is instantiated inside the DRCF component. Each candidate component has two extra ports. One is a DONE signal port routed to the Configuration Scheduler (CS). The port is used to acknowledge the CS that this task can be safely swapped out. The other is connected to a shared memory that saves the data to be preserved during reconfiguration. The Input Splitter (IS) is an address decoder and it manages all incoming Interface-Method-Calls (IMCs). The CS monitors the operation states of the candidate components and controls the reconfiguration process.

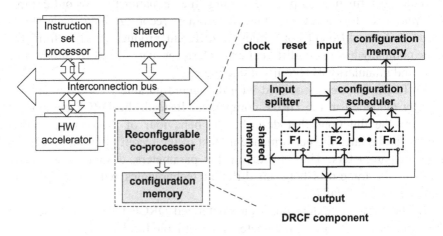

Figure 5-8. System-level Modelling of Reconfigurable SoC.

The DRCF component works as following. When the IS captures an IMC to a candidate component, it will hold the IMC and pass the control to the CS, which decides if reconfiguration is needed. If so, the CS will call a reconfiguration procedure that uses the parameters specified in step 1 to generate memory traffic and associated delays to mimic the reconfiguration latency. After the CS finishes the reconfiguration loading, the IS will dispatch the IMC to the target module. If the module cannot be activated at the moment, a message of request to reconfigure the target module will be put into a FIFO queue and the IMC will return with the value of FALSE. When a module finishes its operation, it will send a DONE signal to the CS, and the CS will check if there is any waiting message in the FIFO queue. If so and it is possible to activate the waiting module, the CS will call the reconfiguration procedure.

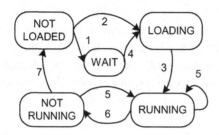

State Definitions:

NOT LOADED:	module is only in the configuration memory
LOADING:	module is being loaded
WAIT:	module is waiting in a FIFO queue to be loaded
RUNNING:	module is running
NOT RUNNING:	module is loaded, but not running

State Transition Conditions:

1. IMC to the module occurs & not enough resources
2. IMC to the module occurs & enough resources
3. CS finishes the loading
4. Other modules finish & enough resources
5. IMC to the module occurs
6. Module finishes
7. CS flushes the module

Figure 5-9. Reconfiguration state diagram.

The context switching with pre-emption is a common approach in operating systems, the implementation of which is easy due to the regularity of the register organization. In the DRCF component, the pre-emption of a running module is not supported, since it would require a very costly implementation of the hardware module in order to store the internal registers and states of the module.

The modelling method is for non-blocking IMCs. The method supports the use of blocking, but the system bus will be blocked when a called candidate component is not loaded and unblocked when the reconfiguration is done. The reason is to maintain synchronization between the SW initiators and the candidate components.

While this is a generic description of the context switching process, designers can use different CS models when candidate components are mapped to different types of reconfigurable devices, such as partial reconfiguration and single-context device. The auto-transformer, which is presented in the following sections, uses a context switching mechanism for single-context devices.

There is a state diagram common to each of the candidate components. Based on the state information, the CS makes reconfiguration decisions for all incoming IMCs and DONE signals. A state diagram of partial reconfiguration is presented in Figure 5-9. For single context and multi-context reconfigurable resources, similar state diagrams can be used in the model. The main advantage of the modelling method is that the rest of the system and the candidate components need not to be changed between a static approach and run-time reconfiguration approaches, which makes this method very useful in making fast design space exploration.

5.3   Automatic Transformer for SystemC Based Extensions

The DRCF transformer is a tool that can automatically transform the SystemC code of a static system to the SystemC code of a reconfigurable system. It takes two inputs. One is SystemC models of the initial architecture, and another is a script file that specifies which modules should be moved into the DRCF component and all the other relative information, e.g. parameters for the DRCF template.

Outputs of the program are the modified architecture as well as SystemC models of the DRCF component and the memory associated with it. A Makefile for compilation is an optional output.

A UML diagram of the static structure of the DRCF transformer is shown in Figure 5-10.

For the sake of brevity, operations and attributes of classes are ignored in the diagram. The transformer uses Opencxx [11] as the basic C++ parser to analyse the SystemC code. The *ClassHandle* and *SystemCClassHandle* manage the analysed information. A lex&bison-based parser is developed to read the user script file and the results are stored using class *DRCFReqInfo*. The class *DRCFTemplateHandle* is responsible for generated the SystemC models of the DRCF component and the memory block associated with it. Finally, *DRCF_driver* is the kernel that controls the process of transformation.

The flow of the transformation is shown in Figure 5-11. In the first phase, each module that is a candidate component to be implemented in reconfigurable hardware is analyzed. The used bus interface and the bus ports are analyzed so that the DRCF component can implement the same interfaces and ports. After modules are analyzed, the transformer moves to analyze each instance of the modules in architecture. Firstly, the declaration of each instance is located and then the constructors are located and copied to a temporary database. When all instances are analyzed, the DRCF component is created from a DRCF template.

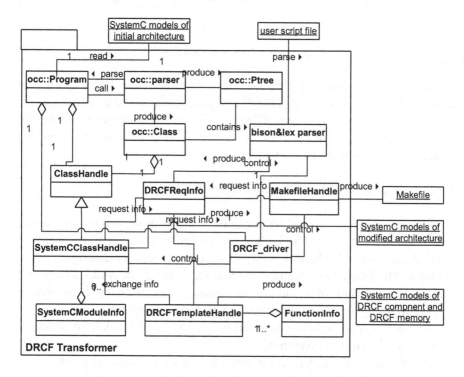

Figure 5-10. Software specification of DRCF transformer in UML.

The ports and interfaces analyzed in the first phase are inserted to the DRCF template and then the component to be implemented in dynamically reconfigurable hardware is instantiated according to the declaration and constructor located in the second phase. The DRCF template contains a context scheduler to mimic the context switching process, an input splitter that routes data transfers to correct instances, and instrumentation processes.

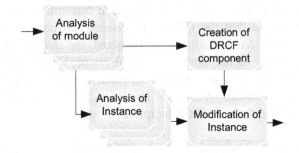

Figure 5-11. Transformation flow.

During simulation, data related to reconfiguration latency will be automatically captured by the DRCF component and saved in a text file for analysis. A VCD (Value Change Dump) file will also be produced by the DRCF component, so the configuration effect can be analysed via standard waveform viewers that can read VCD format file.

5.3.1   Example of the Transformation Process

A simple example of what will be done to the SystemC modules is shown next. The initial static system includes three hardware accelerators, *hwacc1*, *hwacc2*, and *hwacc3*. There is no direct control dependence among the three modules. The estimation results show that the *hwacc2* and *hwacc3* consume about equal amount of resources as *hwacc1*. The decision is to assign *hwacc1* to one context, and the other two to a second context.

A fragment of code below is a part of the hardware accelerator, *hwacc1*, which is modelled using SystemC.

```
class hwacc1: public sc_module, public bus_slv_if
{
  public:
      sc_in_clk clk;
      sc_port<bus_mst_if> mst_port;
      ...
};
```

In the first phase of operation, the ports and interfaces of the module are analyzed. In this case, the module implements one interface *bus_slv_if*, which is the slave interface of a bus; the module has two ports *clk* and *mst_port*, which represent the clock input and the master interface of a bus. Next, the top-level module is analyzed to understand the structure of the system. The code below shows the instantiation of the module in a hierarchical module named *top*.

```
SC_MODULE(top)
{
  sc_in_clk clk;
  hwacc1 *hwa;
  hwacc2 *hwb;
  hwacc3 *hwc;
  bus *system_bus;
  SC_CTOR(top){
      system_bus = new bus("bus");
      system_bus->clk(clk);
  /* signal bindings for hwacc1 */
      hwa = new hwacc("HWA",HWA_START,HWA_END);
      hwa->clk(clk);
      hwa->mst_port(*system_bus);
      system_bus->slv_port(*hwa);
  /* signal bindings for hwacc2 */
      hwb = new hwacc("HWB",HWB_START,HWB_END);
      hwb->clk(clk);
      hwb->mst_port(*system_bus);
      system_bus->slv_port(*hwb);
  /* signal bindings for hwacc3 */
      hwc = new hwacc("HWC",HWC_START,HWC_END);
      hwc->clk(clk);
      hwc->mst_port(*system_bus);
      system_bus->slv_port(*hwc);
  }
};
```

After the analysis of the top-level module, the declarations, constructors, the port bindings and the interface bindings in terms of the module *hwacc1*, *hwacc2*, and *hwacc3* are removed. This hierarchical module is then updated to use the DRCF component instead of the hardware accelerators. The modified code is shown below. Notice that the declaration, the constructor and the bindings are modified for a new instance of *drcf*.

```
SC_MODULE(top)
{
  sc_in_clk clk;
  drcf *drcf_inst_1;
  bus *system_bus;
  SC_CTOR(top){
      system_bus = new bus("bus");
      system_bus->clk(clk);
      drcf_inst_1 = new drcf("DRCF1");
```

```
        drcf_inst_1->clk(clk);
        drcf_inst_1->mst_port(*system_bus);
        system_bus->slv_port(*drcf_inst_1);
    }
};
```

The actual DRCF component created from the DRCF template is shown in the code below. In the code, the declarations, constructors and the interface bindings of the hardware accelerators are copied from the original top-level module. The port bindings are automatically modified. The text that is in italics is the code that was dynamically created from the information saved for the instances of the modules *hwacc1*, *hwacc2*, and *hwacc3*. What was already in the template is the *arb_and_instr()* method that handles the context scheduling and instrumentation. The instrumentation is a SystemC process that keeps track of the configuration status.

```
    class drcf: public sc_module, public bus_slv_if
    {
      public:
        sc_in_clk clk;
        sc_port<bus_mst_if> mst_port;
        hwacc1* hwa;
        hwacc2* hwb;
        hwacc3* hwc;
        SC_HAS_PROCESS(drcf);
        void arb_and_instr();
        sc_uint<ADDW> get_low_addr();
        sc_uint<ADDW> get_high_addr();
        bool read(...);
        bool write(...);
        SC_CTOR(drcf){
          SC_THREAD(arb_and_instr);
          sensitive_pos<<clk;
      /* signal bindings for hwacc1 */
          hwa = new hwacc("HWA",HWA_START,HWA_END);
          hwa->clk(clk);
          hwa->mst_port(*mst_port);
      /* signal bindings for hwacc2 */
          hwb = new hwacc("HWB",HWB_START,HWB_END);
          hwb->clk(clk);
          hwb->mst_port(*mst_port);
      /* signal bindings for hwacc3 */
          hwc = new hwacc("HWC",HWC_START,HWC_END);
          hwc->clk(clk);
```

```
hwc->mst_port(*mst_port);
ContextInfo* cont0 = new ContextInfo(...);
cont0->insert_module(hwa);
ContextInfo* cont1 = new ContextInfo(...);
cont1->insert_module(hwb);
cont2->insert_module(hwc);
contexts.push_back(cont0);
contexts.push_back(cont1);
  }
};
```

6.   USING WORKLOAD MODELS FOR DESIGN SPACE EXPLORATION

As reconfigurability adds a new dimension to the design space, a reliable method of analyzing performance of the resulting system is needed. An architecture that fits well to the application at hand avoids many design problems in the later detailed design stages, but it is difficult to find the bottlenecks in the architecture early enough.

Traditionally, models of system or sub-system start with a purely behavioural description which contains only the functionality to be performed. Then, the models are gradually refined towards a certain type of implementation, and concrete information is inserted into models in each refinement. However, when designers have an initial architecture in mind at the beginning of design, the performance simulation of it cannot be performed until each model is iteratively refined to a level of abstraction, which contains enough low-level information from the architecture point of view. The traditional modelling method does not only delay the performance simulation of the architecture, but it also makes difficult the exploration of design space in terms of looking for alternative architectures.

Using SystemC as a system modelling language provides the opportunity to perform architecture-space exploration in the early phase of design. This is achieved using SystemC transaction-level workload operation models. The workload model separates the computation and communication. At the transaction level, the load of computation is represented using timed information either cycle accurate or not, and load of communication is represented using combined factors, such as type of transaction, bandwidth of bus, latency of accessing memory, behaviour of bus arbiter and so on. The timing information of computation could be the estimate from a supporting tool, such as the estimation approach introduced in the preceding section, or from designers' experience. The factors related to the communication are architecture-dependent and could be set as parameters. Thus, by tuning these parameters in performance simulation, the best architecture in terms of certain performance aspect could be easily found.

The following code is a simple example that models the computation latency of a hardware accelerator. The macro *DELAY_CYCLES* can be given a different value when the task is mapped to a different kind of processing unit.

```
void accelerator::do_process(){
    remain_im = accu_im & mask_coeff;
    /* delay DELAY_CYCLES cycles until write */
    for(delay=0;delay<DELAY_CYCLES;delay++){
      wait();
    }
    if(remain_re>=64 && accu_re>0){
      data_out_short_real = accu_re/128 +1;
    }
}
```

The following code presents transaction-level communication from a master to a memory block through a system bus. The delay associated with the bus arbitration process is irrelevant to the model of the master block.

```
void accelerator::do_process(){
    bus_port->read(mem_in_addr,&data_in);
    unpack(data_in, data_in_real, data_in_imag);
    rot_re[i1] = data_in_real.to_int();
    rot_im[i1] = data_in_imag.to_int();
}
```

At the Architecture Definition the best architecture has to be searched using iterative design and modelling. System-level performance simulations can be performed by building workload models of the application in order to simulate them on candidate architectures. The developed SystemC architecture and workload modelling and simulation approach is depicted in Figure 5-12.

It operates on transaction-level of abstraction. The simulation results are estimates of computational complexity of each block, estimates of communication and data storage requirements, and characteristics of the architecture and the mapped workload.

The workload model at transaction level contains information about how long each processing stage takes and how it communicates with other processes. The communication can be modelled using SystemC resources such as ports, interfaces and channels. The initial architecture is derived from the application analysis results. To get the full benefit of this modelling scheme, utilization of each resource of the architecture will be measured in terms of e.g. idle cycles, data waiting cycles and operation cycles.

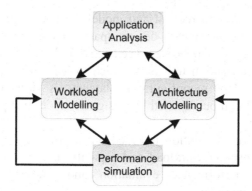

Figure 5-12. Principle of transaction-level performance simulation.

Workload models are used to generate load on the architecture by mapping the functional operations to processing elements. Communication and synchronization of processes and processing elements has been implemented using part of the memory as communication registers.

The workload and architecture models can be refined during simulations. Some system characteristics and load effects can easily be adjusted by modifying such parameters as clock frequencies, bus widths, latencies of memory operations and speed-up factors that can be used to model various candidate implementations of parallelism or to model speed-up of hardware accelerator implementations etc. These refinements are continued until the resource utilization rates that are acceptable for the application are reached.

7. CONCLUSIONS

The extensions to the SystemC for supporting the design of SoCs incorporating reconfigurable parts are described in this chapter. The extensions are based on standard features of the SystemC 2.0.

SystemC encapsulates C/C++ descriptions of algorithms into an implementation neutral system model by exploiting either standard or user defined communication mechanisms, e.g. different types of channels.

One extension is the methods and prototype tool support for the estimation of software execution time on an ISP and hardware execution time and resource consumption on an FPGA, which provides information for system partitioning and selection of candidate components for reconfigurable design. Traditional analysis methods and tools are required in a full SW/HW system partitioning.

Another extension is the DRCF modelling method that can automatically detect the reconfiguration request and model the reconfiguration overhead. This technique allows for fast design space exploration, since explored modules can be easily switched between fixed and reconfigurable modules. A prototype transformation tool is provided to help to generate the DRCF SystemC model. The reconfiguration latency is derived from a few parameters, which can be adjusted by designers in design space exploration step.

The system-level simulation is based on the transaction-level SystemC model and uses abstract workload and capacity models of application and architecture for performance evaluation and studying of alternative architectures and mappings.

The main benefit of the extended SystemC based approach is that it enables modelling and performance evaluation of a system containing reconfigurable parts already at the system level before devoting efforts to the detailed and implementation design.

REFERENCES

1. S. Swan (2001) An Introduction to System Level Modeling in SystemC 2.0. http://www.systemc.org
2. SystemC (2002) Functional Specification for SystemC 2.0, Update for SystemC 2.0.1, version 2.0-Q. April 5. http://www.systemc.org/
3. K. Tiensyrjä, M. Cupak, K. Masselos, M. Pettissalo, K. Potamianos, Y. Qu, L. Rynders, G. Vanmeerbeeck and Y. Zhang (2004) SystemC and OCAPI-XL Based System-Level Design for Reconfigurable Systems-on-Chip. Forum on Specification & Design Languages (FDL 2004). 14-17 September 2004. ECSI, Grenoble, France, pp. 428-439
4. Y. Qu and J.-P. Soininen (2003) Estimating the utilization of embedded FPGA co-processor. Euromicro Symposium on Digital Systems Design, 2003 (DSD 2003), pp. 214–221
5. A. Pelkonen, K. Masselos and M. Cupak (2003) System-level modeling of dynamically reconfigurable hardware with SystemC. The 17th International Parallel and Distributed Processing Symposium (IPDPS 2003), pp. 174–181
6. Y. Qu, K. Tiensyrjä and K. Masselos K (2004) System-level modeling of dynamically reconfigurable co-processors. Proceedings of the 14th International Conference on FPL (LNCS 3203), pp. 881-885
7. R. P. Wilson, R. S. French, C. S. Wilson, S. P. Amarasinghe, J. M. Anderson, S. W. K. Tjiang, S. W. Liao, C. W. Tseng, M. W. Hall, M. S. Lam and J. L. Hennessy (1994) SUIF: An Infrastructure for Research on Parallelizing and Optimizing Compilers. Proceedings of the 7th ACM SIGPLAN symposium on Principles and practice of parallel pro programming, pp. 37−48

8. D.D. Gajski, N. Dutt, A. Wu and S. Lin (1997) High-level synthesis: Introduction to chip and system design. Kluwer Academic Publishers, Boston

9. P. G. Paulin and J. P. Knight (1989) Force-Directed Scheduling for the Behavioral Synthesis of ASICs. IEEE Transactions on Computer-Aided Design of Integrated Circuits and Systems 6: 661−679

10. J.-P. Soininen, J. Kreku, Y. Qu and M. Forsell (2002) Mappability estimation approach for processor architecture evaluation. Proceedings of the 20th IEEE Norchip Conference (NORCHIP 2002), pp. 171−176

11. S. Chiba (1998) Open C++ Tutorial. http://opencxx.sourceforge.net

Chapter 6

OCAPI-XL BASED APPROACH

Miroslav Čupák and Luc Rijnders
IMEC, Kapeldreef 75, B-3001 Leuven, Belgium

Abstract: This chapter describes the OCAPI-XL based modelling techniques and tools that support the design of reconfigurable systems-on-chip (SoC). To allow modeling of reconfigurability features at system level, we developed: 1) new software process type in OCAPI-XL, 2) coupling of OCAPI-XL to SystemC for co-simulation, and 3) context switching from one resource towards another (software, reconfigurable hardware).

Key words: Configuration overhead; context switching; design space exploration; dynamic reconfiguration; estimation; mapping; partitioning; reconfigurable; reconfigurability; SystemC; OCAPI-XL; system-on-chip.

1.   INTRODUCTION

Heterogeneous HW/SW systems on a chip (SoC) present one of the vital challenges for design methodologies of today. OCAPI-XL (OXL) is a C++ based design environment for development of concurrent, heterogeneous HW/SW applications. It abstracts away the heterogeneity of the underlying platform through an intermediate-language layer that provides a unified view on SW and HW components. The language is directly embedded in C++ via a creatively designed set of classes and overloaded operators [1], and has an abstraction level between assembler and C.

OXL's design-flow, as depicted in Figure 6-1, starts at high (typically C/C++) level and goes all the way down to the implementation in a sequence of incremental steps.

N.S. Voros and K. Masselos (eds.), System Level Design of Reconfigurable Systems-on-Chips, 133-151.
© 2005 *Springer. Printed in the Netherlands.*

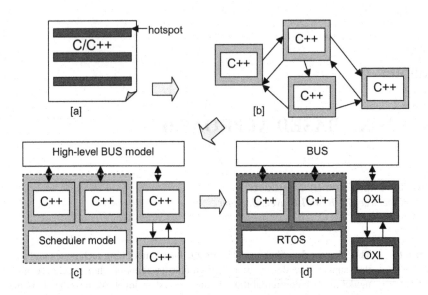

Figure 6-1. OCAPI-XL design-flow: [a] single-threaded C++ specification; [b] parallelisation; [c] modelling of architecture constraints, and [d] refinement to OCAPI-XL embedded language.

The OXL design flow can be divided as follows:

1. Identification of hotspots, i.e. heavily used parts of code, where parallelisation would be beneficial. (Figure 6-1[a]). This can be done using C/C++ tools, such as *quantify* or *gprof*.

2. Partitioning of the single-threaded C/C++ code into parallel tasks using OXL's concurrency and communication primitives (Figure 6-1[b]) based on the analysis from the previous step. The main goal is to get parallel C++/OXL code, functionally equivalent to the single-threaded original.

3. Mapping of the functional model from step 2 onto the architecture described via a set of constraints, like number of SW-processors, relative HW-clock speed, or communication resource sharing (Figure 6-1[c]). This step adheres to the Y-chart modelling approach [2,3] (i.e. strict separation of functionality and architecture).

4. Complete refinement of selected processes to OXL embedded language (Figure 6-1[d]), which can then be used in HW, as well as in SW scenarios.

At the stages from 2 to 4, OXL provides the designer with simulation results as well as quantitative figures of system throughput, activity, performance etc., that is, supplies with an important feedback directing new refinement steps. Additionally, while it allows the designer to stay within the same C++ based framework during the whole design process, it also provides hooks for coupling other simulation engines or environments according to the designers' needs.

2. THREADED PROCESS OCAPI-XL'S EXTENSION

The OXL scheme of embedding a language within C++ removes the need for a completely new language-framework with supporting tools and environment, since one can reuse most of any existing C++ tools. On the other hand, it also creates two genuine problems: how to mix the new language with native C/C++ code and how to translate existing C/C++ code into the new language in an incremental way. These problems were addressed in OXL, the solution was however not general enough and failed for significant category of applications, as it will be shown later. This deficiency is addressed by the presented threaded-process extension. It closes the gap in the OXL design-flow, and provides a conceptually sound, generic link between the high-level C/C++ code and OXL Embedded Language (OXL-EL) also in the cases that were not handled properly by the existing techniques.

The threaded-process extension will be presented as follows: first, we will show the existing technique for integration of C/C++ and OXL code, pinpoint its weakness and indicate the proposed solution. Afterwards, we will describe the implementation of co-simulation library.

2.1 OXL and C/C++ Code Integration

The distinguishing feature of OXL is the language providing a unified semantic model for HW and SW. The OXL language is embedded in C++, which allows easy integration of existing C++ code to OXL. Unfortunately, it also makes the boundary between C++ and OXL code somewhat blurred to the designer, which can be quite dangerous, considering that we have co-existing two semantically different languages. In this section, we will firstly outline the basic idea of the OXL language implementation and its interaction with C++; secondly, we will introduce the original technique for integration of C++ and OXL code: the Foreign-Language Interface (FLI), and pinpoint its deficiency, and finally we will introduce the idea of a threaded process as a way to deal with the FLI deficiencies.

2.1.1   OXL Embedded Language

The notion of class, as a new type semantically equivalent to the predefined ones, is one of the central ideas of C++. It allows to use C++ as a meta-language, where classes represent types of the new language and the class' code defines the new language semantically. Operator overloading adds a bit of syntactic sugar allowing to make such a language closer to C syntax and easier to read or write. This embedded-language approach is used in OXL to implement the unified HW/SW language in a way sketched in Figure 6-2. The so-called operator classes like addop represent the constructs of the OXL-EL, and their member functions (like sim()) implement their semantic meaning. These classes do not directly execute their code, but rather create a runtime structure called heap, similar concept to a byte-code, which is later interpreted during simulation or code-generation.

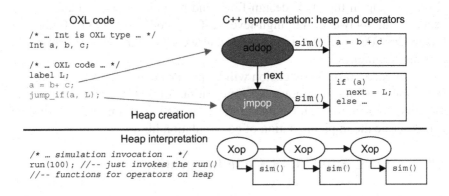

Figure 6-2. C++ representation of the OXL language.

Thus, C++ and OXL code can not be combined without any restriction. The original method of combining C++ and OXL code was the FLI mechanism.

2.1.2   Foreign-Language Interface Mechanism

The FLI mechanism is the original interface for integration of C++ code into OXL-EL. Conceptually, an FLI is just another operator class of the OXL-EL, but without its functionality fixed. Rather, it can be specified by the designer via overriding of the virtual run() function of the fli class. Consequently, an FLI object is seen from OXL simulation engine as a single, atomic instruction into which inputs are passed, and from which outputs are read (Figure 6-3). There is no limitation of the amount of the code within the `run()` function. One can read a file or a socket, write to a terminal, communicate with another system process, etc. The possibilities are without limits [4] – almost.

The FLI mechanism has an important limitation: for OXL, the code within the `fli::run()` function must be a single, atomic instruction. This limitation prohibits usage of other OXL instructions (e.g., message read or semaphore post) in the code. We found it acceptable for data-dominant applications, where large parts of code behave like an atomic instruction so that they can easily be split into a few FLI classes. These FLI objects can then be manipulated by the designer at will, e.g., grouped into processes, annotated with timing info, mixed with refined code, etc.

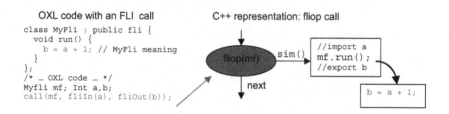

Figure 6-3. OXL representation and simulation of an FLI object.

In the case of control-oriented applications, where the potential atomically executable parts of code are much smaller, the FLI technique becomes quite controversial, since it typically results in either a few big FLI objects, or many small ones. In the first case, no OXL primitives can be used inside the `run()` method so that all OXL-EL features are unavailable to the designer. In the second case, the code atomisation requires a lot of tedious, error-prone work leading to a refined code, which is hard to read and maintain. To deal with such cases, an alternative mechanism for the combination of unrefined C++ and OXL-EL was needed.

2.1.3 Threaded Processes Extension: Idea

OXL-EL has its own way to deal with concurrency via heap structure and operator classes. Of course, it is not sufficient to implement processes with native C/C++ thread of control. In order to introduce such processes into OXL, we must devise a way of extending the OXL kernel with some thread-library primitives, since only such primitives can provide the support for handling of arbitrary C/C++ code concurrently. Additionally, the extension must preserve OXL kernel's complete control of the simulation to ensure its correctness, regardless whether the thread library is pre-emptive or not. The ultimate goal of such an extension it to provide a conceptual way for the combination of plain C/C++ code and OXL concurrency primitives without the need to refine every line of code into OXL-EL. It must allow one to use the full set of OXL communication and synchronization primitives inside the threaded process so that the high-level modelling features of OXL can be used. Its basic idea can be demonstrated in Figure 6-4, where the original C++ code is split into two OXL threaded processes, communication via OXL communication primitives wrapped in C functions. The splitting of C++ code in the case of threaded processes needs only be driven by parallelization requirements, and not from the code-atomicity demand of the FLIs.

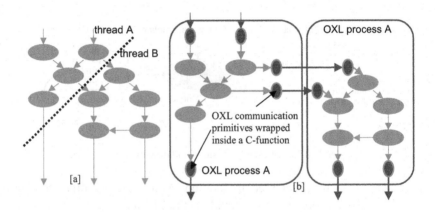

Figure 6-4. Threaded processes from user's point of view.

2.2   Thread Process Extension: Implementation Requirements

There are, in principle, three essential requirements for the implementation of threaded-process extension:

1. Ease of use: ideally, the FLI-like API (i.e. requiring to derive from a class and override a virtual function) should be provided.
2. Backward compatibility: should not influence any existing OXL code. Ideally the extension should be implemented as a plug-in.
3. No restriction on the C/C++ code within the threaded processes (unless imposed directly by the underlying thread library).

Next sections describe, how SystemC library can be used for implementation of the threaded process extension compliant with the above-mentioned requirements.

3.   SYSTEMC IMPLEMENTATION OF OCAPI-XL THREADED PROCESS EXTENSION

SystemC (SC), in principle, provides an implementation of a thread library bundled together with an event-driven simulation engine with notion of virtual time. Thus SC can be used for implementation of the threaded-process extension with the additional bonus of automatically having an OXL/SC co-simulation environment. Such an environment brings together the advantages of OXL (architecture-modelling features, or OXL-EL), and SC (de-facto-standard modelling environment with a significant tool support). There is one additional requirement for the implementation (on top of the three presented in the previous section): the underlying code should only use the standardized API of SC and avoid any implementation-specific features (like, the scheduler call-back function present in the OSCI reference implementation) to make the ensuring OXL/SC environment portable.

The basic structure of the OXL/SC co-simulation environment is shown in Figure 6-5.

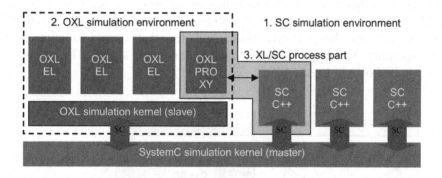

Figure 6-5. OXL/SC co-simulation environment structure.

It consists of three domains within the SC environment:
1. The native SC processes controlled only by the SC simulation kernel.
2. The OXL domain running in a single SC thread and controlled by the OXL kernel acting as a slave to the SC simulation kernel.
3. XL/SC processes (i.e. equivalents of the previously described threaded processes) running C++/SC code with access to OXL's communication and synchronization primitives. These processes contain internally two parts: the SC part running the C++ or SC code, and a small OXL proxy process, which is active when the process is executing an OXL synchronization/communication primitive.

3.1   OXL Environment within SC

The essential idea of the common OXL/SC environment is to let the whole OXL part run in a single SC thread and to use SC synchronization mechanisms inside a modified OXL simulation kernel to synchronize with the rest of the (SC) world. Since the OXL kernel is single-threaded (the concurrency behaviour is implemented via the heap and operator classes), there are no thread compatibility problems created by this setup.

3.1.1   XL/SC Process Implementation

The implementation of the threaded SC/XL process is the crucial part in the SC/XL link. Internally, it consists of two co-operating entities: OXL proxy process and SC thread. The underlying idea is rather straightforward: the process is executing either C++/SC code in SC domain, or OXL concurrency primitive in the OXL domain. In the first case, it is controlled by the SC simulation kernel and is of no interest for the OXL simulation kernel except for its local time (to avoid possibility of anti-causal simulation when OXL scheduler advances its time). In the second case, it is inside the OXL environment, executing a communication/synchronization statement requested from the C++/SC code, and for that period of time the corresponding SC thread is simply blocked inside SC kernel. The complete scheme is shown in Figure 6-6.

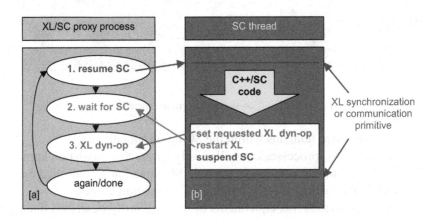

Figure 6-6. XL/SC process implementation via [a] OXL proxy process and [b] SC thread.

The interaction between OXL proxy and SC-thread parts deserves a closer look. The proxy process uses three new operator classes introduced in this co-simulation library:

1. First one, capable to restart the corresponding SC thread (Figure 6-6[a1]).
2. Second one, waiting for the response from SC-thread keeping the proxy process blocked inside the OXL domain (Figure 6-6[a2]).
3. Third one, whose contents can be dynamically changed according to the operation requested from the SC-thread (Figure 6-6[a3]), performs the currently set operation inside the OXL domain.

With these new operators, it is possible to define the OXL proxy process interacting with the SC thread as shown in Figure 6-6[a]. When we look at the OXL proxy process/SC-thread interaction from the SC side, it goes over following sequence of steps:

1. The SC-thread starts in a suspended state and waits till it is not resumed from the OXL proxy process.
2. After resumption, the SC-thread runs until it ends, or a function invoking an OXL communication/synchronization primitive is called. In that case, it updates the dynamic operator in the proxy process with information about the requested XL primitive operation, restarts the proxy process and suspends itself, until awakened again from the OXL proxy process. As a side note: blocking and resumption of a SC thread can be achieved via simple dynamic event waiting/notification scheme available in SC 2.0 [5].

Finally, the presented scheme also requires come changes to the OXL scheduler, since it has to take into account XL/SC processes.

3.1.2 OXL Slave Scheduler

The OXL scheduler must be slightly modified so it could take into account the fact that the XL/SC processes can be outside of its control at certain moments during the simulation. To account for that, the event-dispatch loop of the OXL scheduler may only be allowed to advance in time (i.e., dispatch event with a higher time-stamp than the one of the last dispatched event), if either no processes are in SC domain, or their local time is at least equal to the time of the to-be-dispatched OXL event (otherwise, an anti-causal simulation may happen, after some the XL/SC processes would return into OXL domain with an older time-stamp than event-dispatcher). A pseudo-code of the modified event-dispatch loop is shown in Figure 6-7. Also the new scheduler must hold some additional bookkeeping data about the currently running XL/SC processes.

```
                                                    original OXL event-dispatch loop code
void XLScheduler::ev_loop() {
  while(true) {
    if(event_queue_XL.empty()) {
      if (procs_SC.empty()) break; //-- event queue empty and no procs in SC
      else wait(wake_up); //-- else wait till an SC process returns to XL
    else {
      if(next_time_XL == time_SC) {//-- next XL event has current time-stamp?
        event_queue_XL.dispatch(); //-- we can dispatch it, since no event
                                   //-- with a smaller time can be unprocessed
      } else { wait(wake_up, next_time_XL - time_SC); }
    }
  }
}
```

Figure 6-7. Modified OXL event-dispatch loop.

3.2 **Alternative Threaded Process Implementations**

SC is only one of possible thread environments suitable for implementation of the *threaded process* extension. Essentially any threaded library can be used, employing a similar implementation strategy, i.e. C++ code running inside threads and controlled from proxy processes within OXL environment. Also, the modified OXL scheduler can be simpler in such a case, since it does not have to run as a slave to other simulation kernel. We have successfully implement the *threaded process* extension with *pthread* and *GNU pth* libraries on various operating systems.

4. **SOFTWARE PROCESSES SCHEDULING EXTENSION**

Performance of real life software is highly dependent on the operating system it is running on. Especially, if multi-thread or multi-process software is considered, the influence of the operating system's scheduler is highly influencing the overall performance. Since in OXL the system is described in a parallel communicating processes model, modelling software scheduling will benefit the accuracy of the software performance model.

In the high-level software model of computation (procHLSW) concurrency is considered at the processor level. This means that for every process there is a separate processor assumed (see Figure 6-8).

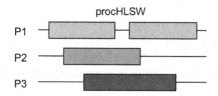

Figure 6-8. High-level software behavior over time.

Naturally, in real life this will typically not be the case. In realistic software implementation there will be an operating system that allows all the processes to be assigned to the same software processing resource. So from the performance point of view the processes are not at all running concurrently, but they are being sequentialized by the operating system scheduler onto the processing unit. To model such behaviour in the OXL performance model, a separate process type having this behaviour has been introduced: `procManagedSW` (see Figure 6-9).

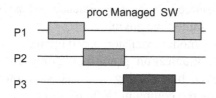

Figure 6-9. Sequentializing computation over time.

To be able to create a process of the type `procManagedSW` the designer must first create a scheduling object. This scheduler will perform the actual sequentialisation of all the processes that will be attached to this object. The way this is done is defined in one of the member methods of this scheduling object. Currently a simple round-robin scheduler, a scheduler with scheduling priorities and a priority scheduling with aging effect are provided. Additionally, user can define its own scheduling objects to model the behaviour of the scheduler present in the target operating system. OXL assumes a non-pre-emptive scheduler, so it is up to the processes to hand over control to the operating system. This can be done by either blocking on a communication primitive of by allowing a context switch (by calling the `sync_()` call).

It is important to realize that switching between the different SW tasks is not penalty-free. It always takes certain number of time (and especially for reconfigurable architectures) to change form one to another task. In order to come to most accurate performance results, context-switching overhead has to be considered in a performance model. This feature has been added to the OXL environment. User can define for every process created, extra context switching time (as an argument of the scheduler object, or using extra setcsoverhead() method), which is then applied to that process during the OXL simulation.

5.   BUS MODELING EXTENSION

The OXL high-level model consists out of concurrent tasks communicating with semaphores and/or message queues. Both have non-blocking write (semaphore unlock, message send) and a blocking read (semaphore lock, message receive) accesses. When doing high-level system modelling, the focus lies more in the functional correctness rather than the correct behaviour in the time domain. At this level, all communication channels are usually considered in parallel and without delay. Depending on the targeted architecture, some of these channels can be mapped onto a shared communication resource. As a consequence, transfers on these channels cannot occur at the same time anymore.

This section explains a bus model extension based upon distinct properties of processes types and communication primitives, going from high-level communication features over bus sharing and access protocols onto a complete C++ model for a shared communication resource.

5.1   Modelling Bus Sharing

Connecting of software processor and the hardware by means of a certain bus structure has always an implication on the performance of the system. A bus can usually not be shared by different tasks at the same time. So it is necessary to adapt the model in such a way that bus-sharing properties come into play.

Bus sharing is very similar to task scheduling. In both cases one resource (either processor or the bus) needs to be shared. This means that accesses (processing or data transfers) will have to occur sequentially in time. We could construct a first model for the bus based upon the already present task scheduling properties. By introducing scheduled dummy processes onto the bus channels we obtain the required behaviour.

Our initial bus model consisted out of the bus writer and bus arbiter parts (see Figure 6-10).

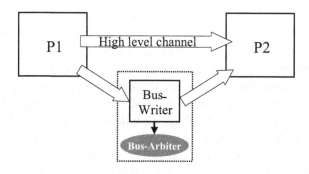

Figure 6-10. Initial Bus Model.

5.1.1   Bus Writer Tasks

A process manager will schedule tasks to ensure that at each moment maximum one writer has access to the bus. In order to reuse the already present scheduling properties, these tasks are software-like in behaviour. By not annotating any operations within these tasks, they are executed in virtual zero time. By annotating a single bus access with fixed time duration, bus transfer delays can be taken into account.

5.1.2   Bus Arbiter

A bus arbiter is part of the communication resource itself. It is responsible for deciding which task is allowed to communicate via bus. How this decision is done differs from one bus architecture to another. Also, depending on the type of bus, transfers could be a single value, or it could mean transfer of a whole burst or values. It may very well be that the type of request issued to the arbiter influences the final decision.

The bus arbiter has a lot of similarities with a task scheduler, and to construct our model, we will actually use such a (native) construct to build our bus model. Since in our methodology the user is allowed to define his own process manager, or scheduler, the new one can be created that acts like the real target bus arbiter.

5.2   Modelling Bus Access Protocol

To make our model even more accurate, we could replace the fixed time annotation of one transfer with more exact value. In an actual transfer it is the bus access protocol that makes that a transfer requires some time. But this access protocol may be different depending again on the type of transfer requested. In a burst-transfer it will most-probably not be necessary to include the overhead of the bus-request/acknowledge nor the time needed to reverse the direction of the bus (come out of tri-state).

These particularities will most probably not have such a great impact on the overall performance if this fixed timing annotation is properly chosen, but out of consistency, we could include the protocol in our model as well.

To include the bus access protocol, we will add an additional task to either the writer side, the reader side or at both sides. These tasks will model the protocol for each transfer. Since these protocols are usually specified as hardware access schemes, the protocol task(s) will be of the hardware type. The final bus model including protocol modelling is graphically represented in Figure 6-11.

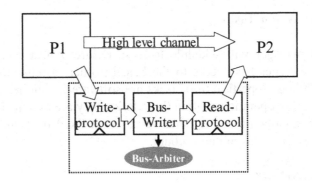

Figure 6-11. Final bus model block diagram.

6.   HIGH-LEVEL MODELLING OF CONTEXT SWITCHING

When designing in OCAPI-XL, application code can be assigned to the following process types:

- A high level abstraction for (scheduled) software targets (procHLSW, procManagedSW).
- Two abstractions for creation of ANSI-C software (procANSIC, procMTHRC).
- A high level abstraction for hardware targets (procHLHW).
- An abstraction for FSMD hardware targets based on OCAPI 1.0. (procOCAPI1).
- A high level abstraction for integrating with SystemC (procSC).

Assigning code to a process affects its simulation, inter-process communication and also code generation, which is the final step when heading for an implementation of the design. At the early stages of the project, the user usually works on simulations to obtain correct OCAPI-XL simulation results of the system. At this stage, the code is assigned to high level process types (procHLSW, procHLHW, procManagedSW). Later, the code is refined towards implementation targets, being either SW or HW. During the refinement step, high-level SW processes have to be rewritten to procANSIC or procMTHRC types. This is to allow single threaded ANSI-C code generation for procANSIC types and multi threaded ANSI-C code generation for procMTHRC processes.

Processes, which target HW implementation, must first be refined to the procOCAPI1 process type. Subsequently, a HDL code (VHDL, Verilog) is generated for each procOCAPI1 process.

6.1 Reconfigurable Context Switching Process

For reconfigurable processes, we consider relocating tasks from the reconfigurable logic to the ISP and vice versa. Therefore, the reconfigurable processes should be specified both as SW and HW processes, as they can potentially be relocated to a different resource at run-time.

The simulation model for task relocation described in the next subsection supports high-level simulation of HW and SW processes, with an opportunity of performance estimation, which takes the reconfiguration time into account.

After refinement, HW and SW code generation can be done for each reconfigurable process, so that the process can be started either as HW or as SW. The code to transfer state information is not automatically generated, and has still to be inserted explicitly by the designer when dynamic reconfiguration with state information (context) memory is needed.

In OCAPI-XL, no parsing of code is done. Source code (ANSI-C for SW targets, HDL for HW targets) is created for the parts of the system model where the OCAPI-XL objects are used. Generation of SW and HW implementations of communication primitives is also supported.

6.2 Simulation Model for Task Relocation

The ability to (re)schedule a task either in hardware or software is an important asset in a reconfigurable systems-on-chip. To support this feature, a possible (high-level) implementation and management of hardware/software relocatable tasks in OXL have been investigated.

The proposed solution uses a high level abstraction of the task state information. The entire relocation process is illustrated in Fig. 6-12. In order to relocate a task, the OS can send a switch message to that task, at any time

(1). Whenever that signalled task reaches a switch point, it goes into an interrupted state (2). In this state, all the relevant state information of that switch point is transferred to the OS (3). Consequently, the OS will relocate that task to another processor. The task will be able to re-initialise itself using the state information it receives from the operating system (4). The task resumes by continuing execution in the corresponding switch point (5). It should be noted that a task can contain multiple switch points and that the state information can be different for every switch point. Furthermore, it is up to the application designer to implement the (application dependent) switch point(s) in such a way that the state information that needs to be transferred is minimal.

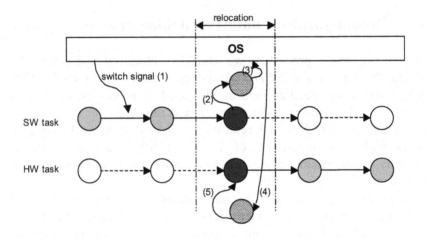

Figure 6-12. Illustration of task relocation.

By modelling task re-scheduling, the application designer can verify on beforehand what the impact is for his particular system. And whether the system performance improvement is not affected too much by the re-location overhead, for example. However, the support for task relocation is not so straightforward as one might expect. Several problems are rising, especially if this task comes from a shared software processing resource (or task scheduler object).

In this case the relocation of the task not only affects the behaviour of the task itself, but also affects the behaviour of the task scheduler was running on, and thus affects all the tasks being scheduled by that scheduling object.

The OCAPI-XL methodology allows keeping track of a bunch of statistics during simulation. By relocating a task, these statistics suddenly get a different meaning. Therefore these statistics must be corrected so they keep the same meaning.

A first simulation model for task relocation was developed in OXL. The OXL code below illustrates the example of coding context switching for a task P1, switching between the different contexts (procHLHW, procManagedSW), and simulating its behaviour.

```
procDRCF P1("P1");
    //-- initial context: High-Level HardWare (default period of 10)
    P1.context(HLHW);
    //-- second context: software under Round-Robin scheduler(RR)
    P1.context(ManagedSW, &RR);
    //-- next context: High-Level HardWare with period of 2
    P1.context(HLHW, 2);
{

    //-- here goes "normal" OCAPI-XL task code

    //-- upon this operator the task will switch itself
    //to the next context
    switchpoint();

    //-- here goes some more task code

}

    //-- and run the simulation for 2000 cycles
run(2000);
```

Within this model, it is in principle also possible to model resource sharing at the hardware level, by replacing one task with another on the same physical reconfigurable space and adding the appropriate contexts.

Prior to starting the simulation, for each relocatable task all the context information, meaning the processing resources the task will run on, and there sequence must be known. If this sequence is known, a new operator, called switchpoint, forces the task to be relocated from the current processing resource towards the next processing resource. This task moving also implies that all the statistics of the current context are finalized, and the statistics of the new context are initialised.

7. CONCLUSIONS

The extensions to the SystemC for supporting the design of SoCs incorporating reconfigurable parts are described in this chapter.

Extension of the OCAPI-XL methodology towards introduction of the thread-level library as well as SystemC implementation of OCAPI-XL threaded process are primarily related to system specification and system-level simulation steps. For system specification step it provides the opportunity to consider not only C/C++ but also SystemC code, which becomes the standard specification language used in the industry. Mixtures of the C/C++/SystemC specifications are also possible. For system-level simulation step, the extension broadens the application scope to control dominated applications, which were not possible to simulate with the existing OCAPI-XL library. Secondarily, this extension affects also the mapping step by providing novel communication and synchronization primitives used within the threaded processes. Although the thread-level library and SystemC implementation of OCAPI-XL threaded process extension has been developed in context of methodology for reconfigurable SoCs, it is generally applicable to all kind of designs.

Software process scheduling extension enhances the system-level simulation step by providing a sequentialisation of software processes. As recent reconfigurable architectures offer one or more processors implemented as soft cores or embedded processors inside the reconfigurable fabric, it is necessary to extend the software process execution modelling on these processors in OCAPI-XL. By introducing this extension, OCAPI-XL is able to provide the means of modelling for both sequential and parallel execution of the processes.

Extension of OCAPI-XL by bus model is related to system-level simulation and mapping steps of the design methodology. As reconfigurable architectures often use dedicated bus communication schemes, modelling of the bus behaviour provides the means for early performance estimation during system-level simulation. The set of bus modelling related primitives introduced to OCAPI-XL provide sufficient means for expressing the system-level bus behaviour.

In order to cover one of the distinguishing features of reconfigurable architectures, modelling of dynamic reconfiguration is implemented in OCAPI-XL library. This includes providing new process characterized by the ability to represent different contexts (different process types) and alternate them during high-level simulation. Inserting switch points specified by the designer in the high-level specification does alternation of the processes during simulation. This way the system-level simulation models the dynamic reconfiguration of selected processes in early stage of the design and provides feedback about influence of different dynamic reconfiguration schemes on performance of the system.

REFERENCES

1. OCAPI-XL manual. IMEC's internal document
2. Kienhuis B et al (1997) An approach for quantitative analysis of application-specific dataflow architectures. Proc IEEE Int Conf On Application-Specific Syst Arch and Proc:338−349
3. Vanmeerbeeck G (2001) et al Hardware/software partitioning of embedded system in OCAPI-XL. Proceedings of the 9th International Symposium on Hardware / Software Codesign − CODES:30−35
4. Pasko R at al (2000) Functional verification of an embedded network component by co-simulation with a real network. IEEE International High Level Design Validation and Test Workshop − HLDVT:64−67
5. SystemC v2.0 manual. http://www.systemc.org

PART C

DESIGN CASES

Chapter 7

MPEG-4 VIDEO DECODER

Miroslav Čupák and Luc Rijnders
IMEC, Kapeldreef 75, B-3001 Leuven, Belgium

Abstract: The OCAPI-XL based approach was applied in the MPEG-4 video decoder design case with aim to validate the system level reconfigurability extensions on a typical multimedia application. The MPEG-4 case represents a scenario where tasks are relocated between software and reconfigurable hardware depending on the level of quality of service requested by the user. The MPEG-4 video decoder has been implemented on Xilinx Virtex-II Pro multimedia demonstration platform.

Key words: Design space exploration; static reconfiguration; estimation; mapping; partitioning; reconfigurable; reconfigurability; OCAPI-XL; system-on-chip.

1. MPEG-4 VIDEO DECODER IN A NUTSHELL

Next-generation of mobile multimedia devices will provide a rich array of digital video and multimedia applications to enhance the end-user experience. MPEG-1 and MPEG-2, the first two video standards from the Moving Pictures Experts Group (MPEG), were fundamental in creating widespread acceptance of digital video formats. Their successor, MPEG-4, can be considered as the first true multimedia standard, taking an object based approach for the coding and representation of natural or synthetic audiovisual content [1,2,3]. It offers a flexible toolset, adaptable to a large variety of requirements, while interoperability among different terminals is guaranteed. The bitrates start at a few hundred bits for synthetic audio up to hundreds of Mbps for the modelling and description of complex multimedia scenes.

The MPEG-4 natural visual decoder (video decoder) is a block-based algorithm exploiting temporal and spatial redundancy in subsequent frames. It takes as input a bitstream, a sequence of bits representing the coded video

N.S. Voros and K. Masselos (eds.), System Level Design of Reconfigurable Systems-on-Chips, 155-177.
© 2005 *Springer. Printed in the Netherlands.*

sequences, compliant with the ISO/IEC 14496-2 standard [4]. The bitstream starts with identifying the visual object as a video object (other kinds, like still textures exist). This video object can be coded in multiple layers (scalability). One layer consists of Visual Object Planes (VOPs), time instances of a visual object (i.e. frame). A decompressed VOP is represented as a group of MacroBlocks (MBs). Each MB contains six blocks of 8 x 8 pixels: 4 luminance (Y), 1 chrominance red (Cr) and 1 chrominance blue (Cb) blocks. Figure 7-1 defines the macroblock structure in 4:2:0 format (the chrominance components are downsampled in horizontal and vertical direction) [5].

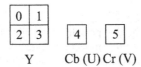

Y   Cb (U) Cr (V)

Figure 7-1. 4:2:0 Macroblock structure.

Two compression techniques are discriminated. In the intra case, the MB or VOP is coded on itself using an algorithm that reduces the spatial redundancy. Inter coding relates a macroblock of the current VOP to MBs of previously reconstructed VOPs and reduces in this way the temporal redundancy.

Figure 7-2 presents the structure of a simple profile video decoder, supporting rectangular I and P VOPs. An I VOP or intra coded VOP contains only independent texture information (only intra MBs). A P-VOP or predictive coded VOP is coded using motion compensated prediction from the previous P or I VOP, it can contain intra or inter MBs. Reconstructing a P VOP implies adding a motion compensated VOP and a texture decoded error VOP. Note that all macroblocks must be intra refreshed periodically to avoid the accumulation of numerical errors. This intra refresh can be implemented asynchronously among macroblocks.

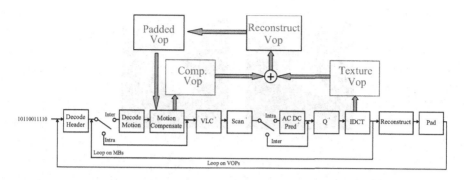

Figure 7-2. The data flow of the simple profile decoder.

1.1   Motion Compensation

A video sequence typically has a high temporal correlation between similar locations in neighbouring images (VOPs). Inter coding (or predictive coding) tracks the position of a macroblock from VOP to VOP to reduce the temporal redundancy. Figure 7-3. follows the movement of a hand and feet of a dancer in the successive frames. The motion estimation process tries to locate the corresponding macroblocks among VOPs. MPEG-4 only supports the translatory motion model.

Figure 7-3. Temporal correlation in a video sequence.

The top left corner pixel coordinates (x,y), specify the location of a macroblock. The search is restricted to a region around the original location of the MB in the current picture, maximally this search area consists of 9 MBs (illustrated in Figure 7-4). With $(x+u, y+v)$, the location of the best matching block in the reference, the motion vector equals to (u,v). In backward motion estimation, the reference VOP is situated in time before the current VOP, opposed to forward motion estimation where the reference VOP comes later in time.

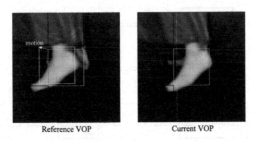

Reference VOP   Current VOP

Figure 7-4. Motion estimation process.

As the true VOP-to-VOP displacements are unrelated to the sampling grid, a prediction at a finer resolution can improve the compression. MPEG-4 allows motion vectors with half pixel accuracy, estimated through interpolation of the reference VOP. Such vectors are called half pel motion vectors.

A macroblock of a P VOP is only inter coded if an acceptable match in the reference VOP was found by the motion estimation (else, it is intra coded). Motion compensation uses the motion vector to locate the related macroblock in the previously reconstructed VOP. The prediction error $e(x,y,t)$, the difference between the related macroblock $MB(x+u, y+v, t-1)$ and the current macroblock $MB(x,y,t)$ is coded using the texture algorithm.

$$e(x,y,t) = MB(x,y,t) - MB(x+u, y+v, t-1)$$

Reconstructing an inter MB implies decoding of the motion vector, motion compensation, decoding the error and finally adding the motion compensated and the error MB to obtain the reconstructed macroblock.

1.2   Texture Decoding Process

The texture decoding process is block-based and comprises four steps: Variable Length Decoding (VLD), inverse scan, inverse DC & AC prediction, inverse quantization and an Inverse Discrete Cosine Transform (IDCT). Except for the IDCT, all blocks have to produce numerical identical results to ISO/IEC 14496-2 and ISO/IEC 14496-5.

The VLD algorithm extracts code words from Huffman tables, resulting in a 8x8 array of quantized DCT coefficients. Then, the inverse scan reorganizes the positions of those coefficients in the block. In case of an intra macroblock, inverse DC & AC prediction adds the prediction value of the surrounding blocks to the obtained value. This is followed by saturation in the range [-2048, 2047]. Note that this saturation is unnecessary for an inter MB. Because no DC & AC prediction is used, the inter MB DCT coefficients are immediately in the correct range.

Inverse quantization, basically a scalar multiplication by the quantizer step size, yields the reconstructed DCT coefficients. These coefficients are saturated in the range $[-2^{bitsPerPixel+3}, 2^{bitsPerPixel+3}-1]$. In the final step, the IDCT transforms the coefficients to the spatial domain and outputs the reconstructed block. These values are saturated in the range $[-2^{bitsPerPixel}, 2^{bitsPerPixel}-1]$.

1.3 Error Resilience

The use of variable length coding makes the (video) bitstreams particularly sensitive to channel errors. A loss of bits typically leads to an incorrect number of bits being VLC decoded and causes loss of synchronization. Moreover, the location where the error is detected is not the same as where the error occurs. Once an error occurs, all data until the next resynchronisation point has to be discarded. The amount of lost data can be minimized through the use of error resilience tools: resynchronisation markers, data partitioning, header extension and reversible variable length codes [4].

2. IMPLEMENTATION PLATFORM

The following sections describe the families of Virtex-II boards, clarify the concept of embedded soft cores that can be build inside of the Virtex-II FPGAs and explain the process of selection the suitable platform for MPEG-4 video decoder demonstrator.

2.1 Virtex II Boards

Xilinx offers a line of prototype boards for the Virtex-II series of Platform FPGAs. These boards are intended to provide testing and modeling of the design functionality. They come with documentation, cables, connectors, and a set of reference designs intended to gain knowledge about efficient use of the boards. An example of family of Virtex-II boards is the Xilinx Multimedia Development Board (XMDB). The board is designed to be used as a platform for developing multimedia applications. The board supports PAL and NTSC television input and output, true color SVGA output, and an audio codec with power amplifier, as well as Ethernet and RS-232 interfaces. Several push button and DIP switches are available for user interaction with the system. The embedded SystemACE™ controller allows for high-speed FPGA configuration from CompactFlash™ storage devices. Figure 7-5 shows the Xilinx Multimedia Development Board components.

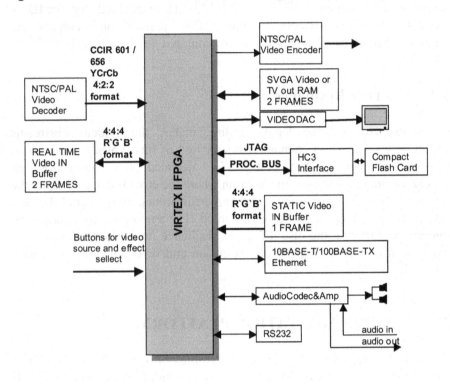

Figure 7-5. Xilinx Multimedia Demonstration Board.

As the Virtex-II family doesn't contain any embedded SW core on the chip, the usual solution for the designs where processor core is demanded is the use of virtual processor that is created out of bits of code in reconfigurable fabric.

2.2   MicroBlaze soft processor

The MicroBlaze [6] is a virtual microprocessor that is built by combining blocks of code called cores. It uses Harvard-style RISC architecture with separate 32-bit instruction and data buses running at full speed to execute programs and access data out of on-chip or external memory (see Figure 7-6). The core contains 800 lookup tables and 32 general purpose registers with three-operand instruction format. Its standard peripheral set is designed to work with IBM's CoreConnect on-chip bus to simplify core integration. The MicroBlaze pipeline is a parallel pipeline, divided into three stages: Fetch, Decode, and Execute.

In general, each stage takes one clock cycle to complete. Consequently, it takes three clock cycles (ignoring delays or stalls) for the instruction to complete. Each stage is active on each clock cycle so three instructions can be executed simultaneously, one at each of the three pipeline stages.

MicroBlaze runs theoretically at 150 MHz and delivers 123 D-MIPS.

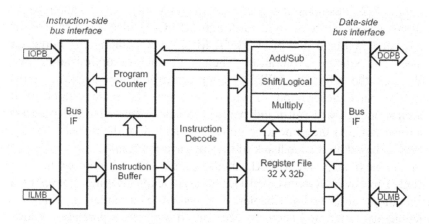

Figure 7-6. A view on a MicroBlaze processor.

2.3 Platform Selection

On any SoC project, the overall goal of the platform selection process should be to find the platform that reduces risk as much as possible. The considered risk involves both technical and non-technical aspects that have to be taken into account. From the proposed methodology point of view, it should be possible to implement its design features (described at Chapter 4) at every level of the design. At system-level design phase, it has to be determined which design components are possible to implement or reuse on the platform and determine how they interact. At detailed design phase, when the design is refined, it must be guaranteed that all the refinements are supported by the platform. At the same time proper platform verification strategy has to be known to prove that solutions are not based upon incorrect assumptions. The implementation phase then involves building the system at the specified platform. Here the important factors in fulfilling the design goals are: possible IP reuse of the components, experience of the design team with board environment, vendor support, etc.

Taking all this aspects into account, Xilinx Virtex-II Multimedia Demonstration Board with embedded MicroBlaze soft processor core has been selected as the most appropriate demonstration platform for MPEG-4 video decoder. XMDB board provides sufficient support for implementation of the MPEG-4 demonstrator case.

3. MPEG-4 VIDEO DECODER DESIGN FLOW

Figure 7-7 illustrates the proposed methodology (see Chapter 4) and positions associated tools used to design MPEG-4 video decoder. The first design step involved use of ATOMIUM [7] methodology for initial data transfer and storage exploration. Then a number of optimizations were applied on the reference MPEG-4 simple profile video decoder, starting from the basic description of video decoder. The main output of the optimization was a platform independent, memory optimized video decoder described in C. At the same time initial architecture of the decoder has been proposed, based on the feedback of the optimization results.

In the next step, OCAPI-XL methodology (see Chapter 6) has been exploited to cover System-Level Design and Detailed Design phases of the methodology described at Chapter 4. A system level model, in which both functionality and architecture can be described separately, allowed performance modelling at a high level of abstraction. A refinement strategy and executable specifications at all levels enabled a structured path towards implementation. At the end of the design flow, OCAPI-XL generated VHDL code for the reconfigurable HW parts of the design.

The implementation phase of the decoder design has been fully covered by the commercial tools dedicated for optimal implementation of the SW and HW parts on the selected platform. Synplify-Pro FPGA and CPLD synthesis tool has been used for implementing the FPGA part of the design. SW implementation, co-verification and board integration has been supported by the Xilinx's ISE and EDK toolsets.

4. SOFTWARE VERSION OF THE MPEG-4 VIDEO DECODER

This section covers the initial analysis of the MPEG-4 video decoder, decoder optimizations and describes pure SW version of the decoder.

4.1 Definition of the Functionality Testbench and Decoder Pruning

The Verification Model (VM) software used as input specification was the FDIS (Final Draft International Standard) natural visual part [3]. Having working code at the start of the design process can overrule the tedious task to implement a system from scratch. Unfortunately, the software specification was very large and contained many different coding styles of often varying quality.

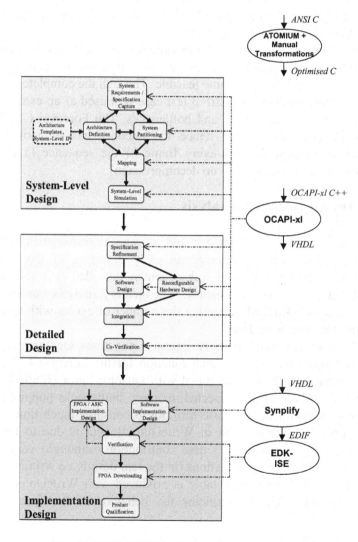

Figure 7-7. MPEG-4 Video Decoder Design and Tool Flow.

Moreover, the VM contained all the possible MPEG-4 decoding functionality (i.e. of all possible profiles) resulting in oversized C code distributed over many files. The video decoder on itself has a code size of 93 files (.h and.c source code files) containing 52928 lines (without counting the comment lines).

A necessary first step in the design was extracting the part of the reference code corresponding to the desired MPEG-4 functionality of the given profile and level. ATOMIUM pruning [7] was used to automate this error-prone and tedious task. It removed the unused functions and their calls based on the instrumentation data of a testbench representative for the desired functionality. This implied careful selection of the set of input stimuli, which has to exercise all the required functionality.

Applying automatic pruning with this functionality testbench reduced the code to 40% of its original size. From this point, further manual code reorganization and rewriting became feasible. Through the complete analysis and optimizations, the Foreman CIF 3 test case was used as an example for the detailed study of the effects and bottlenecks. The Foreman CIF 3 test case uses no rate control and hence the decoder has to activate the decompression functionality for every frame of the sequence (a skipped frame just requires displaying but no decompression).

4.2   Initial Decoder Analysis

An analysis of the data transfer and storage characteristics and the computational load initially allowed an early detection of the possible implementation bottlenecks and subsequently provided a reference to measure the effects of the optimizations. The memory analysis was based on the feedback of ATOMIUM. Counting the number of cycles with Quantify assessed the computational load.

Table 7-1 lists the most memory intensive functions together with the relative execution time spent in each function for the Foreman CIF 3 test case. The timing results were obtained with Quantify on a HP9000/K460, 180 MHz RISC platform. As expected, memory bottlenecks popping up at this platform independent level also turn out to consume much time on the RISC platform. The time spend in WriteOutputImage is due to network overhead and disk accessing. It's time contribution (although) very large, was neglected during the optimizations (in the real design, no writing to disk occurs). The last column of the table is produced with WriteOutputImage disabled. The following list explains the behavior of the functions in Table 7-1:

- **VopMotionCompensate**: Picks the MB positioned by the motion vectors from the previous reconstructed VOP. In case of halfpell motion vectors, interpolation is required.
- **BlockIDCT**: Inverse Discrete Cosine Transform of an 8 x 8 block.
- **VopTextureUpdate**: Add the motion compensated and texture VOP.
- **BlockDequantization**: Inverse quantization of the DCT coefficients.
- **CloneVop**: Copies data of current to previous reconstructed VOP by duplicating it.
- **VopPadding**: Add a border to previous reconstructed VOP to allow motion vectors to point out of the VOP.
- **WriteOutputImage**: Write the previous reconstructed VOP (without border) to the output files.

Only the IDCT is a computationally intensive function, all the others mainly involve data transfer and storage. The motion compensation and block IDCT together cause more than 40% of the total number of memory accesses, making them the main implementation bottlenecks. Hence, the focus was on these functions during the memory optimizations (i.e. reduce the number of accesses).

Table 7-1. Motion compensation and the IDCT are the memory bottlenecks of the decoder (Foreman CIF 3 test case)

Function name	#accesses/ frame (10^6 accesses/frame)	Relative # accesses (%)	Relative time (%), to disk	Relative time (%), not to disk
VopMotionCompensate	3.9	25.4	16.9	38.34
BlockIDCT	2.8	18.0	9.4	21.25
VopTextureUpdate	1.7	10.7	3.1	6.8
BlockDequantization	0.5	3.0	2.0	4.5
CloneVop	1.2	7.5	1.5	3.46
VopPadding	1.1	7.0	1.4	3.08
WriteOutputImage	1.0	6.2	54.9	-
Subtotal	11.6	74.7	89.1	77.43
Total	15.5	100.0	100.0	100.0

Both for HW and for SW, the size of the accessed memory plays an important role. Accesses to smaller memories have a better locality and hence typically result in a higher cache hit chance for SW and in lower power consumption for HW. Figure 7-8 groups the accesses to 4 memory sizes: frame memory with as minimal size the height width of the VOP, large buffer containing more than 64 elements, buffer with 9 to 63 elements and registers with maximally 8 elements. In this initial analysis stage, the word length of the elements is not considered. 50 % of the total number of accesses is to frame memory, 13 % to a large buffer, 23 % to a buffer and 13

% to registers. As accesses to large memories are most inefficient, the optimizations focused on reducing the accesses to those memories.

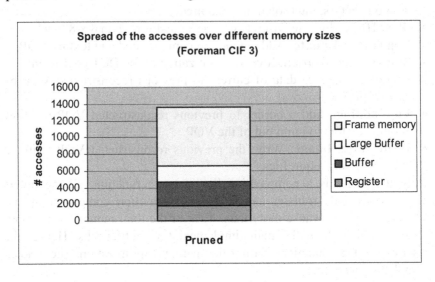

Figure 7-8. Most accesses of the reference decoder are to (large) frame memories.

From the initial analysis of the (pruned) FDIS code, a high-level data flow model has been derived. For every VOP, the algorithm loops over the MBs. First, the motion information is reconstructed. In case of an inter MB, the motion vector is decoded and the motion compensated MB is stored at the current position in the compensated VOP. In case of an intra MB, the compensated MB is stored as all zeros. Secondly, the texture information is decoded. Inverse VLC and inverse scan yield the DCT coefficients. In case of an intra MB, also inverse DC & AC (if enabled) prediction has to be performed. Inverse quantization and IDCT produce the texture MB that is stored at the current position in the texture VOP. When all MBs of the VOP are processed, the reconstructed VOP is composed by adding the compensated and texture VOP. This complete VOP is copied as it is needed at the next time instance for the motion compensation as reference. Finally, a border is added to this reference VOP to allow the motion vector to point out of the image. The resulting VOP is called the padded VOP. This illustrates that the data exchanged between the main parts of the decoder is of frame size. Hence the data flow of the reference decoder is VOP based.

4.3 Decoder Optimizations

Decoder optimizations were performed in two phases. During the first phase, the data flow was transformed from frame-based to macroblock-based. In the second phase, a block-based data flow was introduced. These optimizations aimed at the reduction of the number of accesses and the improvement of the locality of data.

The effect of the platform independent optimisations has been assessed by ATOMIUM and has been validated towards software and hardware implementation. The global number of accesses was reduced with a factor 5.4 to 18.6, depending on the complexity of the sequence. The peak memory usage dropped from some megabytes to a few kilobytes. The performance measure showed a consistent speed up. The highest speedup was measured on a PC platform, where the speed up factor varies between 5.9 and 23.5. The proposed architecture contains a single processor and a three level memory organization. The obtained results are generic and allow a rapid evaluation of alternative memory hierarchies.

4.4 Evaluation

After analysis and optimizations of the code, the SW only version of SW MPEG-4 video decoder has been implemented on Xilinx Virtex-II Multimedia Demonstration Board running fully on MicroBlaze soft processor. Board measurements have shown that decoder runs at 0.5 frames per second for a typical CIF video sequence. Even if the SW related acceleration techniques would be used, which would bring a yield of magnitude speed up, no real-time behavior would be achieved. HW acceleration was the only solution to solve the problem. This moves the critical functionality to the FPGA fabric. The HW/SW partitioning of video decoder, as described in the following section, allows for parallel processing in SW and HW, assuming that the time previously consumed by critical blocks is minimal when moved to HW. This way also a communication overhead is reduced.

5. HARDWARE ACCELERATED VERSION OF MPEG-4 VIDEO DECODER

Straightforward implementation of pure SW version of MPEG-4 video decoder resulted in insufficient performance of the system. The next step in the design was proposing the acceleration steps to improve the processing speed towards real-time behavior.

5.1   HW/SW Partitioning

Primary candidates for implementing in hardware became the computation/data transfer most dominant blocks – VopMotionCompensate, BlockIDCT, VopTextureUpdate and BlockDequantization (see Table 7-1). Secondarily, moving those blocks to hardware also influenced partitioning of a number of sub-blocks that were involved in transferring the data between the memories and HW/SW. Especially the accessing the data in memories have had an impact on HW/SW partitioning of sub-blocks. These have been put to HW if efficient cycle count saving could have been obtained. The final HW/SW partitioning of the MPEG-4 video decoder is shown in Figure 7-9.

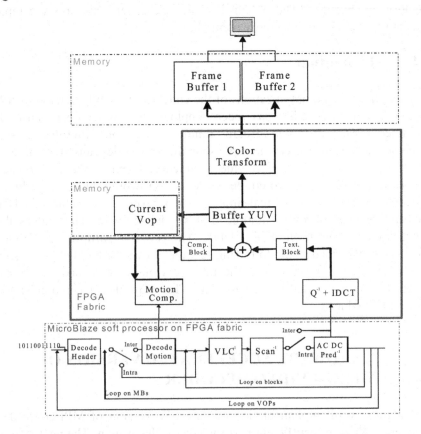

Figure 7-9. HW/SW partitioning for hardware accelerated version of the decoder.

5.2 HW/SW Co-Design in OCAPI-XL

After HW/SW partitioning, the OCAPI-XL model of the accelerated MPEG-4 video decoder has been build (see Figure 7-10). It consists of four main OCAPI-XL blocks: motion compensation (MC), inverse DCT and dequantization (iDCT-Q^{-1}), a process responsible for writing the block to YUV buffer (Block2BufferYUV) and a process which stores the buffered data to current VOP (BufferYUV2CurrentVOP). These blocks further contain smaller OCAPI-XL processes, executing specific function within the block. After introducing proper communication mechanisms between the OCAPI-XL processes, the communication between the MicroBlaze and the HW accelerator have been defined. A memory-mapped interface serves the purpose of communicating between the MicroBlaze and the different HW blocks, both for data and control signals. For memory accesses to YUV buffer and VOP blocks a C++ parametrisable buffer library has been used. To visualize the decoding process, a dual video memory system was proposed for display rendering.

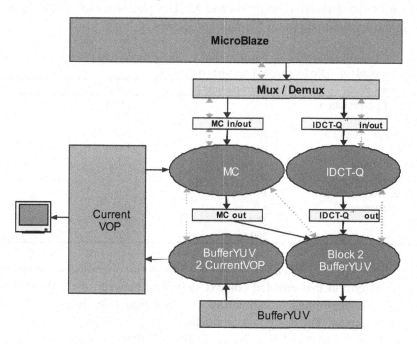

Figure 7-10. HW Accelerator Data and Control Flow.

5.3   Performance Estimation

The next step in the MPEG-4 video decoder design was estimation of the gain that can be obtained by HW acceleration. OCAPI-XL estimation techniques have been involved to solve this task. Prior to performance estimation, the C code for VopMotionCompensate, BlockIDCT, VopTextureUpdate and BlockDequantization blocks have been rewritten to OCAPI-XL processes and refined. We used OCAPI-XL operation set simulation approach for performance estimation, which is analogy to the Instruction Set Simulator approach. The performance models of processors were characterized in a table of operations with associated execution cycle count. The execution cycle counts were obtained from board measurements by execution of small programs.

Operation set approach covers performance estimation of the OCAPI-XL processes. However, as SW part of the decoder was running as a separate thread, it was still necessary to annotate the SW tasks with proper timing information. We have exploited the simulation time results of the pure SW version of the decoder running on the XMDB platform (see Table 7-2) to obtain the approximate times for those processes.

Table 7-2. Time spend in the main functional blocks during the decoding of Foreman CIF 450 kbps, 12 seconds of video for pure SW version on XMDB

Functional Block	Function time (s)	Relative time (%)
Motion Compensation	164.0	28.8
Buffer To Vop	75.8	13.3
VOP Texture Update	105.7	18.6
Q^{-1}/IDCT	105.9	18.6
VLC Decoding	43.7	7.7
Init Block	20.2	3.5
DC AC Reconstruction	14.7	2.6
Read Bitstream	3.0	0.5
Other	36.4	6.4
Total	569.5	100.0

Exploiting OCAPI-XL unified system description, operation set simulator and the extension that enabled context switching between the HW and SW enabled the performance estimation of reconfigurable behavior of the MPEG-4 decoder. The design was modelled in two versions:

- Configured as pure SW version of MPEG-4 decoder, and
- Configured as HW accelerated version.

With a SW processed annotation the performance estimation started with simulation of the pure SW version of the decoder. The frame times have been obtained and the average frame time have been calculated. For pure SW version, the OCAPI-XL processes have been defined of managed SW type, and round-robin scheduler has been exploited. This corresponded with serial execution of processes on a single MicroBlaze processor.

For HW accelerated version of the decoder, the OCAPI-XL processes have been redefined to high-level hardware *(HLHW)* type and the code has been recompiled. HLHW is characterised by detailed modelling between the function and cycle time. This cycle time information has been specified during refinement step in the OCAPI-XL design. Comparing the average frame time with pure SW version, the speed-up of factor 4.2 was estimated.

These two reconfigurable scenarios are switched during the OCAPI-XL simulation at specific switch point inside the OCAPI-XL tasks. It should be noted that switching between the different contexts is fully supported during high-level simulations. However, it is responsibility of the designer to solve the HW/SW task relocation at the implementation level.

5.4 Further optimizations

Estimation of the performance of the HW accelerated decoder indicated improvement of factor 4.2 (frame rate 2.1 fps) compared to the pure software version. Further steps therefore concentrated on improvement the memory access times to obtain real-time performance.

Basic experiments quantified the data transfer const on the multimedia board to allow for assessing the impact of an improved platform. A small C program counted the number of cycles required for a data transfer to the different kinds of available memory on the multimedia board: local memory, block RAM and ZBT RAM (off chip). Table 7-3 lists the different read/write times when the function add/or stack resides in local memory or off chip (ZBT RAM). Table 7-4 measures the amount of cycles spent during the data copy from one kind of memory (row) to another (column).

Table 7-3. Memory access cycles

	Local Stack				Non Local Stack			
	Local Function		Non Local Function		Local Function		Non Local Function	
	Read	Write	Read	Write	Read	Write	Read	Write
Local Memory	6	6	15	15	11	14	28	28
Block RAM	8	7	22	19	13	15	30	32
ZBT RAM	10	7	24	19	15	15	32	32
Stack	4	4	10	10	9	9	20	20

The results in the tables above indicated that cycle savings could be gained by putting the selected object files of the decoder in local memory to make their functions local. Making use of local memories for object code, in connection with pixel packing, the speed up of factor of 2.3 was obtained. Another optimisations included separating MicroBlaze instruction and data busses, Direct Memory Access and exploiting the compiler optimizations. By accumulative applying these performance optimization steps, the frame rate of 25 fps has been obtained. The real-time performance of 30 fps could be easily obtained if the XMDB board maximum clock frequency would not be limited to 81 MHz.

Table 7-4. Data copy cycles

| | Local Stack | | | | | |
| | Local Function | | | Non Local Function | | |
	Local Memory	Block RAM	ZBT RAM	Local Memory	Block RAM	ZBT RAM
Local Memory	8	9	9	20	24	24
Block RAM	10	11	11	27	28	28
ZBT RAM	12	13	13	29	30	30
	Non Local Stack					
	Local Function			Non Local Function		
	Local Memory	Block RAM	ZBT RAM	Local Memory	Block RAM	ZBT RAM
Local Memory	16	17	17	33	37	37
Block RAM	18	19	19	40	41	41
ZBT RAM	20	21	21	42	43	43

5.5   Implementation Details

The video decoder demonstrator was realized on the Xilinx MicroBlaze Development Board, which incorporates Virtex-II xc2v2000 FPGA and embedded MicroBlaze soft processor core. The board is designed to be used as a platform for developing multimedia applications. The board supports five independent banks of 512K x 36bit 130MHz ZBT RAM with byte write capability. This memory is used as video frame buffers store. The embedded SystemACE environment consisting of a CompactFlash storage device and a controller is used for storing the encoded data. The ethernet connection is used to trigger the decoding process form the browser running on neighbouring PC. The decoded sequence is displayed on the monitor connected to the SVGA output of the board.

The MPEG-4 video decoder reads control and configuration settings from file, initializes and starts MPEG-4 decoding. The decoding itself is triggered with a URL request in a browser on a PC connected to the same LAN network as the multimedia board. This request starts up the MPEG-4 Video Decoder on the MicroBlaze which will open and read the control file located in SystemAce flash RAM on the board and then reads the stream of encoded data also stored in flash RAM. Decoder generates YUV frame data that are send to the rendering block and then displayed on the monitor connected to the board.

Resource utilization for the whole MPEG-4 decoding system for xc2v2000 FPGA on the board was 7703 slices i.e. 71% of the resources. The HW accelerator blocks itself consumed 5000 slices, which represents 46% of resources. Allocation data for build-up multipliers, block RAMs and LUTs are shown in Table 7-5.

The clock rate for the decoder was set to a maximum board available 81 MHz.

Table 7-5. Xilinx Virtex-II xc2v2000 FPGA resource utilization

MULT 18X18s	19	33%
RAMB 16s	43	76%
LUTs	11237	52%

6. RESULTS ANALYSIS

6.1 Analysis of Design Methodology Results

6.1.1 Benefits

The key benefit shown on MPEG-4 video decoder is demonstration of ability of high-level simulation-based performance estimation and evaluation of context switching between the different computation resources. Based on the performance estimation results, it is possible to construct trade-off curve for considered HW/SW partitions. This gives the designer opportunity to evaluate at early stage of the design process, which components is beneficial to implement in reconfigurable HW and which ones will be running in SW.

6.1.2 Disadvantages

OCAPI-XL performance estimation is based on the operation set simulation approach. This means that for every operation a true cycle execution has to be provided to get the estimation results as close as possible to a real board execution time. These measurements have to be obtained as a result of board measurements, meaning that the board either have to be available or the estimates of the cycle count are considered. However, ad-hoc estimates introduce a considerable risk that the expected performance results will miss the actual board performance. Therefore, the as exact as possible board measurements are critical to match the high-level simulation with real-life execution on the board.

6.1.3   Summary

Recent platform FPGAs integrate high-performance CPUs within reconfigurable fabric. This combination provides flexible and high-performance system design environment suitable for deployment of a wide variety of applications. Ability of performance estimation and introduction of context switch modelling at the high-level of design brings the advantage of early evaluation of possible reconfigurable HW/SW partitioning decisions. Demonstrating on MPEG-4 video decoder application, the proposed OCAPI-XL based approach has proven the ability to represent a methodology, which successfully copes with reconfigurable SoC design.

6.2   Analysis of Implementation Results

6.2.1   Benefits

Design of MPEG-4 Video Decoder demonstrated, that OCAPI-XL based methodology is highly suitable approach for designing the RSoCs at the system level. It has been shown, that ability of generation of HDL description from refined OCAPI-XL models has an important role with respect to design time. Although the designer is required to put an effort to refine the high-level OCAPI-XL process types to low-level processes, the benefit of fast HDL code generation during the possible design iterations provides overall gain by reducing the HDL re-design and re-simulation time. Moreover, the OCAPI-XL HDL code generator provides the HDL description of communication primitives, interconnection of HDL blocks, HDL testbench generation and easy integration of IP blocks. With respect to the reconfigurability, the automatic HDL code generation is beneficial in flexible generation of different reconfiguration scenarios.

OCAPI-XL dynamically reconfigurable process type (procDRCF) implements context switching between the high-level HW/SW process types. This enables fast exploration of variety of different reconfigurable schemes at high-level design step. The type of the process can be alternated during simulation at arbitrary time, taking into account reconfiguration time overhead. In this sense, the modelling of context switching opens the possibility of modelling of dynamic reconfiguration at high-level of the design.

Having the opportunity of modelling of run-time context switching, designer must have the possibility of performance estimation of different reconfigurations. This allows fast evaluation of high-level decisions and focusing on those parts of the design flow where the gain in performance and efficiency is greatest. Exploiting the OCAPI-XL's Operation Set Simulator approach has shown to be beneficial for annotation of SW processes, implemented on MicroBlaze soft core embedded on FPGA device.

6.2.2 Disadvantages

The experiences from MPEG-4 design indicate that detailed knowledge of the implementation platform is crucial for efficient implementation of the design. Such knowledge can only be obtained from the board experiments, making use small dedicated examples. The examples have to be build to gain the information about the different aspects of the board. Building the testbench examples, obtaining and evaluating results requires allocation of extra design time. For the purpose of MPEG-4 Video Decoder implemented on Xilinx's Multimedia Development Board, the following aspects have been investigated:

- MicroBlaze soft core performance measurements; to investigate the abilities the SW processor.
- Memory explorations; to find out the data transfer times between the different types of memories (local memory, block RAM, off-chip ZBT RAM).
- Vendor support maturity; to investigate the support for newly introduced devices and implementation boards.

As mentioned in the section above, OCAPI-XL based methodology provides full support for high-level modelling of dynamic reconfiguration by introducing context switching. At the implementation level, the situation is much more complicated. Context switching between the HW and SW requires the transitions from one function to another as smooth as possible. This responsibility falls to the real-time operating system (OS), which manages all these complex transitions. Among the other tasks, the OS is responsible for managing the switching between reconfigurable HW and SW on the FPGA, i.e. it must suspend certain tasks that are running so that other tasks can take a turn. To do so, it must remember the state of each task before it stopped execution so that each task can restart from the same state. The implementation of such mechanisms on the recent implementation boards is possible but not straightforward. The true exploitation of dynamic reconfiguration is expected in future platforms.

OCAPI-XL has been extended by a new bus model extension based upon properties of process types and communication primitives. By annotating a bus model with timing information, the bus timing behaviour can be modelled. Although the model could be exploited for high-level bus performance estimation of the MPEG-4 Video Decoder, the decision has been made not to use this approach. Instead, annotation of each type of transfer on the bus by specific timing information obtained from small experiments has been utilized. The main reason for using this approach was insufficient amount of information found in the documentation about the Xilinx's Virtex-II bus architecture, which uses a combination of busses (PLB, OPB) and bridges (PLB2OPB) to communicate between the reconfigurable HW and embedded processor.

6.2.3   Summary

From the descriptions above, the following conclusions can be drawn:
- The system-level OCAPI-XL approach, extended with reconfigurability features, is valid approach for designing RSoCs.
- Dynamic reconfiguration represents implementation obstacle in recent reconfigurable architectures.
- From a designer point of view, deep knowledge of the reconfigurable architectures and platform(s) is still required for efficient mapping of the algorithm.
- There is a lack of implementation information (especially for newly introduced platforms), which can be fed-back to the high-level design phase for accurate high-level modelling.

7.   CONCLUSIONS

The design methodology and flow described in Chapter 4, instantiated for OCAPI-XL, have been used for the realization of MPEG-4 Video Decoder demonstrator targeting high-level performance estimation of the reconfigurable systems. The OCAPI-XL performance estimation techniques, enhanced by new reconfigurable features, demonstrate the ability to obtain high-level statistics, which enable construction of graph with possible shape of curve for best possible partitioning of the application. By modelling the dynamic reconfiguration of selected processes in early stage of the design, feedback about influence of different dynamic reconfiguration schemes on performance of the system is provided. Based on that, the optimal run-time operation of the video decoder application can be selected.

The accuracy of estimations has been measured by comparing the difference between the estimated performance and board performance. The difference is very acceptable 8% for the most relevant test sequence. The MPEG-4 Video Decoder system has been implemented on Virtex-II FPGA with embedded MicroBlaze soft processor core on Xilinx Multimedia Development Board.

REFERENCES

1. MPEG Requirements Subgroup (2001) Overview of the MPEG Standard, ISO/IECJTC1/SC29WG11 N3931
2. JPEG (2004) Available at: http://www.jpeg.org
3. ISO/IEC JTC1/SC29WG11 14496-5 (2000) Information technology— Generic coding of audio-visual objects—Part 5 Amd1: Simulation software N3508
4. ISO/IEC JTC1/SC29WG11 14496-2 (1999) Information technology—Generic coding of audio-visual objects—Part 2: Visual Amendment 1: Visual Extensions N3056
5. Bhaskaran V, Konstantinides K (1997) Image and Video Compression Standards. Algorithms and Architectures, Kluwer Academic Publishers
6. Xilinx (2004), Available at: http://www.xilinx.com/products/design_resources/proc_central/index.htm
7. ATOMIUM (2004), Available at: http://www.imec.be/design/atomium

Chapter 8

PROTOTYPING OF A HIPERLAN/2 RECONFIGURABLE SYSTEM-ON-CHIP

Konstantinos Masselos[1,2] and Nikolaos S. Voros[1]

[1] INTRACOM S.A., Hellenic Telecommunications and Electronics Industry, Greece
[2] Currently with Imperial College of Science Technology and Medicine, United Kingdom

Abstract: In this chapter the prototyping of a reconfigurable System-on-Chip realizing the HIPERLAN/2 WLAN system is discussed. In this case reconfigurable hardware will be exploited to introduce post-fabrication functional upgrades. For the prototyping a commercial platform using components-off-the-shelf has been used. The design flow and system level design methods described in the previous chapters were used for the system development. An evaluation of the design flow and methods in the context of this specific design is also presented.

Key words: Reconfigurable System-on-Chip, prototyping, HIPERLAN/2, functionality upgrading.

1.   INTRODUCTION

In this chapter, the prototyping of a HIPERLAN/2 reconfigurable System-on-Chip on a platform incorporating components-off-the-shelf (COTS) is described. The targeted system has been developed to form the basis for the development of a family of fixed wireless access systems based on HIPERLAN/2 that can be upgraded to support outdoor communications as well. The integration of additional functionality that may be used in a future product improvement (even after product shipment) could rely on the use of software upgrades (this is a commonly used practice in software products). However, due to the expected complexity the system parts that will support the extra functionality (mainly related to complex physical layer DSP tasks) hardware acceleration will be required. The design flow

N.S. Voros and K. Masselos (eds.), System Level Design of Reconfigurable Systems-on-Chips, 179-207.
© 2005 Springer. Printed in the Netherlands.

presented in Chapter 4 has been adopted for the design of the targeted system in order to provide (a) efficient architecture exploration early enough in the design cycle and, (b) a seamless path from specification to implementation.

2. HIPERLAN/2 SYSTEM DESCRIPTION

The HIPERLAN/2 system [1, 2, 3] includes two types of devices: the mobile terminals (MT) and the access points (AP). A typical HIPERLAN/2 architecture is depicted in Figure 8-1. The architectures of the Access Point and the Mobile Terminal are presented in Figure 8-2.

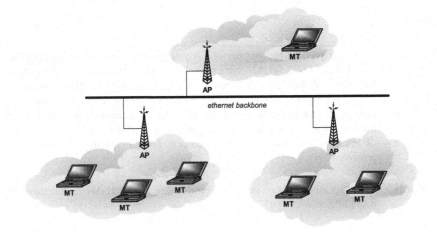

Figure 8-1. Typical HIPERLAN/2 architecture

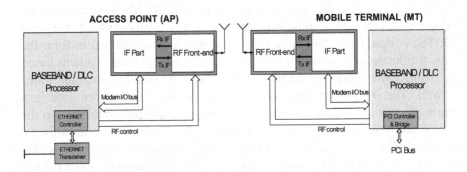

Figure 8-2. Architectures of AP – MT

The HIPERLAN/2 basic protocol stack and its functions are shown in Figure 8-3. The convergence layer (CL) offers a service to the higher layers. The DLC layer consists of the Error Control function (EC), the Medium Access Control function (MAC) and the Radio Link Control function (RLC). It is divided in the data transport functions, located mainly on the right hand side (user plane), and the control functions on the left hand side (control plane). The user data transport function on the right hand side is fed with user data packets from the higher layers via the User Service Access Point (U-SAP). This part contains the Error Control (EC), which performs an ARQ (Automatic Repeat Request) protocol. The DLC protocol operates connection oriented, which is shown by multiple connection end points in the U-SAP. One EC instance is created for each DLC connection. In the case where the higher layer is connection oriented, DLC connections can be created and released dynamically. In the case where the higher layer is connectionless, at least one DLC connection must be set up which handles all user data, since HIPERLAN/2 is purely connection-oriented. The left part contains the Radio Link Control Sublayer (RLC), which delivers a transport service to the DLC Connection Control (DCC), the Radio Resource Control (RRC) and the Association Control Function (ACF). Only the RLC is standardized which defines implicitly the behavior of the DCC, ACF and RRC. One RLC instance needs to be created per MT. The CL on top is also separated in a data transport and a control part. The data transport part provides the adaptation of the user data format to the message format of the DLC layer (DLC SDU). In case of higher layer networks other than ATM, it contains a segmentation and reassembly function. The control part can make use of the control functions in the DLC e.g. when negotiating CL parameters at association time.

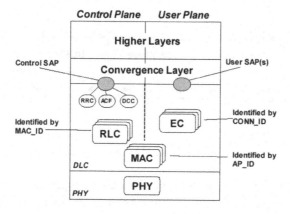

Figure 8-3. HIPERLAN/2 protocol stack and functions

The DLC functions include the following operations:

- (Des)association
- DLC User (de)connection
- encryption, decryption
- (de)framing
- Contention management mechanism
- Broadcast Control Channel (BCCH) and Frame Control Channel (FCCH) analysis and synthesis
- DLC-CL buffering
- Automatic Repeat Request (ARQ) mechanism for asynchronous transactions
- Power Saving
- Dynamic Frequency Selection
- Transmission Power Control

The medium access control (MAC) is a centrally scheduled TDMA/TDD scheme. Centrally scheduled means that the AP/CC controls all transmissions over the air. This is worth for uplink, as well as for downlink and direct mode phase. The basic structure of the air interface generated by the MAC is shown in Figure 8-4. It consists of a sequence of MAC frames of equal length with 2 ms duration. Each MAC frame consists of several phases: Broadcast (BC) phase, Downlink (DL) phase, Uplink (UL) phase, Direct Link Phase (DiL), Random access phase (RA).

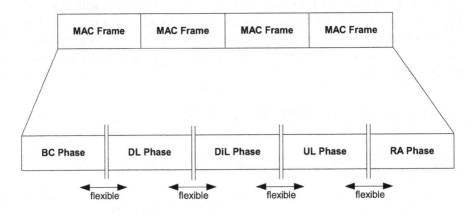

Figure 8-4. Basic MAC frame format

The DL, DiL and UL phases consist of two types of PDUs. The long PDUs have a size of 54 bytes and contain control or user data (see Figure 8-5). The DLC SDU, which is passed from or to the DLC layer via the U-SAP has a length of 49.5 bytes. The remaining 4.5 bytes are used by the DLC for a PDU type field, a sequence number (SN) and a cyclic redundancy check (CRC). The purpose of the CRC is to detect transmission errors and is used, together with the SN, by the EC.

The short PDUs with a size of 9 bytes contain only control data and are always generated by the DLC. They may contain resource requests in the uplink, ARQ messages like acknowledgements and discard messages or RLC information. The same size of 9 bytes is also used in the RCH. The RCH can only carry RLC messages and resource requests. The access method to the RCH is a slotted aloha scheme. This is the only contention-based medium access phase in HIPERLAN/2. The collision resolution is based on a binary backoff procedure, which is controlled by the MTs. The AP/CC can decide dynamically how many RCH slots it provides per MAC frame.

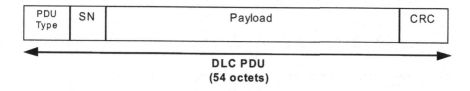

Figure 8-5. Format of the long PDUs

In the physical layer orthogonal frequency division multiplexing (OFDM) has been selected as modulation scheme for HIPERLAN/2 due to its good performance on highly dispersive channels. The channel raster is equal to 20 MHz to provide a reasonable number of channels. In order to avoid unwanted frequency products in implementations the sampling frequency is also chosen equal to 20 MHz at the output of a typically used 64-point IFFT. The obtained subcarrier spacing is 312.5 kHz. In order to facilitate implementation of filters and to achieve sufficient adjacent channel suppression, 52 subcarriers are used per channel, 48 subcarriers carry actual data and 4 subcarriers are pilots which facilitate phase tracking for coherent demodulation. The duration of the cyclic prefix is equal to 800 ns, which is sufficient to enable good performance on channels with (rms) delay spread up to 250 ns (at least).

To correct for subcarriers in deep fades, forward-error correction across the subcarriers is used with variable coding rates, giving coded data rates from 6 up to 54 Mbps. A key feature of the physical layer is to provide several physical layer modes with different coding and modulation schemes, which are selected by link adaptation. BPSK, QPSK and 16QAM are the supported subcarrier modulation schemes. Furthermore, 64QAM can be used in an optional mode. Forward error control is performed by a convolutional code of rate 1/2 and constraint length seven. The further code rates 9/16 and 3/4 are obtained by puncturing. The modes are chosen such that the number of encoder output bits fits to an integer number of OFDM symbols. To additionally accommodate tail bits appropriate dedicated puncturing before the actual code puncturing is applied.

In Table 8-1 the seven physical layer modes are specified, of which the first six are mandatory and the last one based on 64QAM is optional.

Table 8-1. Modes and modulation schemes of HIPERLAN/2

Mode	Modulation	Code rate	Bit rate (Mbps)
1	BPSK	1/2	6
2	BPSK	3/4	9
3	QPSK	1/2	12
4	QPSK	3/4	18
5	16QAM	9/16	27
6	16QAM	3/4	36
7	64QAM	3/4	54

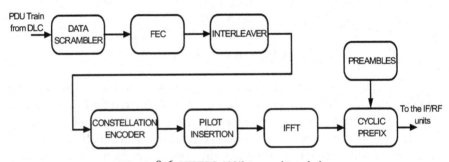

Figure 8-6. HIPERLAN/2 transmitter chain

The transmitter chain of the HIPERLAN/2 physical layer is illustrated in Figure 8-6. In the transmitter path, binary input data are encoded by a standard rate 1/2 convolutional encoder. The rate may be increased by puncturing the coded output bits. After interleaving, the binary values are modulated by using PSK or QAM. The input bits are divided into groups of 1, 2, 4 or 6 bits and converted into complex numbers representing BPSK, QPSK, 16QAM or 64QAM values. To facilitate coherent reception, four pilot values are added to each 48 data values, so a total of 52 values is reached per OFDM symbol, which are modulated onto 52 subcarriers by applying the IFFT. To make the system robust to multipath propagation, a cyclic prefix is added. After this step, the digital output signals can be converted to analog signals, which are then up-converted to the 5 GHz band, amplified and transmitted through an antenna.

The structure and the specifications of the physical layer receiver are not available from the HIPERLAN/2 standard. A generic HIPERLAN/2 receiver is illustrated in Figure 8-7.

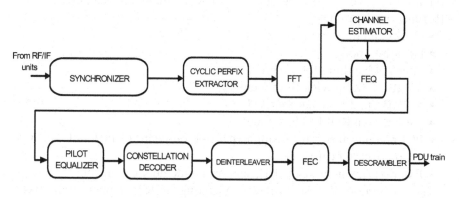

Figure 8-7. HIPERLAN/2 receiver chain

In Table 8-2 the physical layer timing parameters of HIPERLAN/2 system are presented.

Table 8-2. HIPERLAN/2 physical layer timing parameters

PARAMETERS	VALUE
Sampling frequency (f)	20 MHz (T = 50 ns)
Useful symbol part duration (IFFT symbol)	64 x T = 3.2 µs
Cyclic prefix duration	16 x T = 0.8 µs
OFDM Symbol interval	80 x T = 4 µs
Subcarrier spacing	0.3125 MHz (1/3.2 µs)
Spacing between the two outmost subcarriers	16.25 MHz (52 x 0.3125 MHz)
Broadcast burst preamble duration	16 µs
Downlink burst preamble duration	8 µs
Uplink burst short preamble duration	12 µs

3.   IMPLEMENTATION PLATFORM DESCRIPTION

The ARM Integrator/AP AHB ASIC Development Platform has been selected for the prototyping of the HIPERLAN/2 system. The platform is designed for hardware and software development of devices and systems based on ARM cores and the AMBA bus specification.

ARM Integrator supports up to four processors (core modules) to be stacked on the connectors HDRA and HDRB and up to four logic modules to be stacked on the connectors EXPA and EXPB, (a total number of five modules i.e. 2 core modules and 2-3 logic modules are supported). The ARM Integrator provides:

- clocks and three counter/timers
- bus arbitration
- interrupt handling for the processors
- 32MB of 32-bit wide flash memory
- 512KB of 32-bit wide SSRAM
- 256KB boot ROM (8 bits wide)
- PCI bus interface, supporting expansion on-board (3 PCI slots) or in a CompactPCI card rack
- External Bus Interface (EBI), supporting memory expansion.

The Integrator/AP also provides operating system support with flash memory, boot ROM, and input and output resources. Reads from the flash memory, boot ROM, SSRAM, and external bus interface are controlled by the Static Memory Interface (SMI).

3.1   Motherboard architecture

The motherboard hosts the connectors for the core and logic modules that are connected in parallel to the system bus. The block diagram of the motherboard is shown in Figure 8-8. The system controller FPGA provides control functions for the platform (including bus arbitration – up to six masters are supported) and interfaces the core and logic modules (through the system bus) with the rest of the resources on the motherboard (the Flash, SSRAM, ROM, PCI bridge various peripherals – counters, clocks, GPIO, UARTs, keyboard and mouse, LEDs and the interrupt controller).

The system bus is routed between FPGAs on core and logic modules and the AP. This enables the Integrator to support both of the AHB and ASB bus standards. At reset, the FPGAs are programmed with a configuration image stored in a flash memory device. On the AP, the flash contains one image that configures the AP for operation with either an AHB or ASB system bus. On core and logic modules, the flash can contain multiple images so that the module can be configured to support either AHB or ASB.

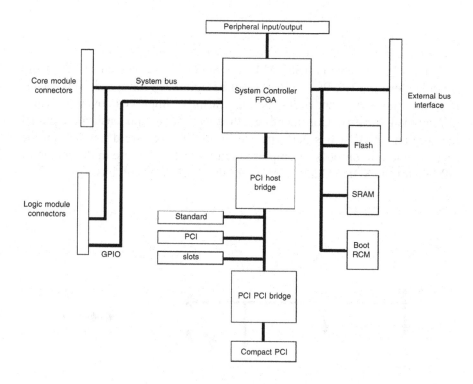

Figure 8-8. ARM Integrator motherboard block diagram

3.2   Core modules

The core module board includes:
- ARM microprocessor chip (ARM7TDMI has been selected)
- 256KB Synchronous SRAM (and relevant controller)
- SDRAM DIMM socket (256MB)
- AMBA system bus interface to platform board
- Clock generators
- Reset controller
- JTAG interface to Multi-ICE™
- Core module FPGA providing system control functions for the core module, enabling it to operate as a standalone development system or attached to a motherboard. The FPGA implements:
    1. SDRAM controller
    2. System Bus Bridge
    3. Reset controller
    4. Interrupt controller

    5. Status, configuration, and interrupt registers
* Multi-ICE, logic analyzer, and optional Trace connectors

The architecture of the core module is shown in Figure 8-9.

The volatile memory (SSRAM and SDRAM) is located on the Core Module close to the CPU, so that it can be optimized for speed. This means that the memory bandwidth is significantly improved over previous development boards. Considerable effort has gone into ensuring optimal memory and AMBA bus performance. Actual figures are dependent on the speed of the microprocessor chip used but typically they are in the region of 50MHz for the SDRAM and 25MHz for the AMBA system bus.

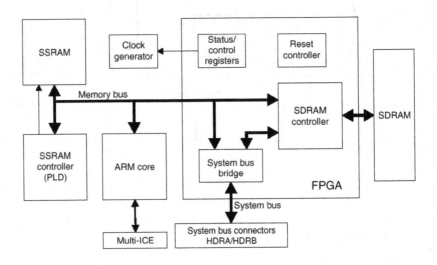

Figure 8-9. Core module architecture

3.3   Logic modules

The logic module comprises the following:
* Altera or Xilinx FPGA
* Configuration PLD and flash memory for storing FPGA configurations
* 1MB ZBT SSRAM
* Clock generators and reset sources
* Switches
* LEDs
* Prototyping grid
* JTAG, Trace, and logic analyzer connectors

•   System bus connectors to a motherboard or other modules

Up to 4 logic modules can be stacked on top of each other, and an Interface Module or an Analyzer Module may be fitted on top of the stack. Core and logic modules handle the interrupt signals differently. Core modules must receive interrupts, but logic modules, that implement peripherals, generate interrupts. The architecture of the logic module is shown in Figure 8-10.

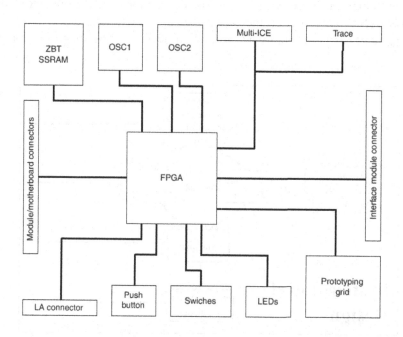

Figure 8-10. Logic module architecture

When used with an Integrator motherboard, the logic modules require a system bus interface. The system bus interface connects the logic module with other Integrator modules. This must be implemented according to the AHB or ASB specifications. The logic module provides the general-purpose interface module connector EXPIM to enable you to add an interface module to the system. The connector provides access to two banks of input/output pins on the FPGA plus a number of control signals.

The logic module provides 1MB of ZBT SSRAM and 4MB of flash memory. A 256Kx32-bit ZBT-SSRAM (Micron part number MT55LC256K32F) is provided with address, data, and control signals routed to the FPGA. The address and data lines to the SSRAM are completely separate from the AMBA buses. This is used for FPGA configuration, and must not be used for any other purpose. Configuration is managed by the configuration PLD.

4.   SYSTEM LEVEL DESIGN

The system level design part of the methodology described in Chapter 4 (and presented in Figure 8-11) has been adopted for the development of the HIPERLAN/2 system. For the system level exploration, the OCAPI-XL environment has been employed (additional details can be found in Chapter 6).

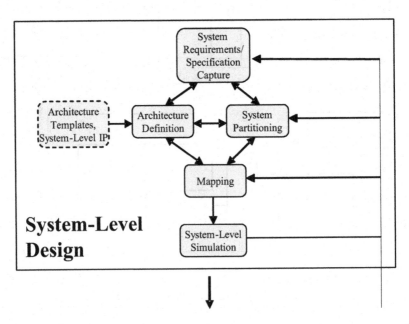

Figure 8-11. System level design part of the proposed methodology

As part of the design process system requirements have been documented, while the targeted functionality has been specified through the development of an executable model. Specifically, an ANSI C model has been developed for the MAC and physical layers' functionality of the HIPERLAN/2 system. The basic structure of the ANSI C model of the targeted functionality is shown in Figure 8-12.

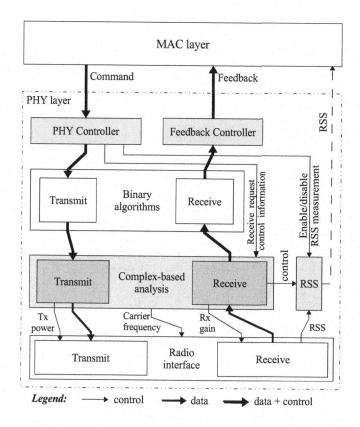

Figure 8-12. Structure of the ANSI C model of the targeted functionality

The physical layer model is divided into two parts: complex numbers based algorithms (mapping, OFDM, PHY bursts) and binary algorithms (scrambling, FEC, interleaving). A block diagram of the physical layer ANSI C model is shown in Figure 8-13. Physical layer submodules are designed as pipelined procedures with unit data processing.

A number of configuration parameters are supported for the physical layer modules:

- width and position of point in fixed point numbers (separate for frequency domain, time domain, FFT calculations, FFT twiddle factors, channel correction and CFO cancellation multipliers)
- number of soft bits in Viterbi algorithm soft value representation
- time synchronization threshold, duration and time-outs
- the highest confidence level threshold of the de-mapper
- sizes of internal buffers (FFT buffers, receiver command buffer, receiver data buffer)

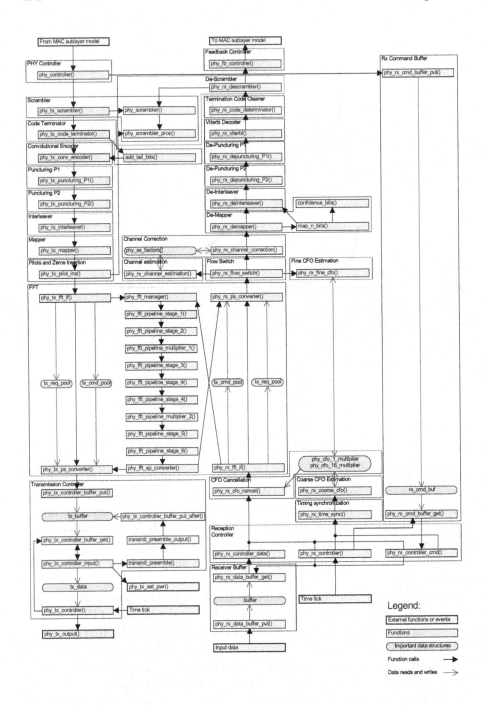

Figure 8-13. Physical layer ANSI C model - major functions and data structures

Physical layer submodules are implemented as procedures, which get as standard parameters: request type, command, command parameters and data. Shared data are represented as global variables. Each submodule has a global variable which value defines a procedure where output will be directed. By default this variable is assigned a value of the procedure corresponding to the next module in physical layer hierarchy. Each physical layer module calls the next one when data portion requested by next module interface is ready. Control information (commands) is forwarded synchronously with data except of shared FFT modules and Viterbi algorithm internals.

Significant part of the high level design of MAC layer is common for Access Point and Mobile Terminal devices. MAC layer high-level design is focused on external interfaces of the sub-layer and its decomposition in cases of Access Point and Mobile Terminal. The block diagrams of the ANSI C models for the Access Point and the Mobile Terminal MAC layers are shown in Figure 8-14 and in Figure 8-15 respectively. In contrast to physical layer MAC modules intercommunication is activated when a logically finished data structure is completely ready. Information is transferred in the form of memory pointers, or copied to some buffer.

Integration testing (simulation) of the ANSI C model has been performed. An environment with one Access Point (AP) and two Mobile Terminals (MTs) has also been emulated. During the emulation, different kinds of traffic have been passed between AP and MTs through an emulated channel in term of fixed point complex numbers.

Analysis of the HIPERLAN/2 computational complexity and performance constraints lead to the allocation of two core modules and two logic modules for the realization of the HIPERLAN/2 system on the ARM Integrator platform. Each core module includes an ARM7TDMI processor and each logic module includes a Xilinx Virtex E 2000 FPGA (0.18µm, 6 metal layers, with 500K usable gates and 832 Kb of additional RAM (BlockRAM) and built-in clock management circuitry (8 DLLs)). Logic and core modules communicate using the bus of the platform (AMBA).

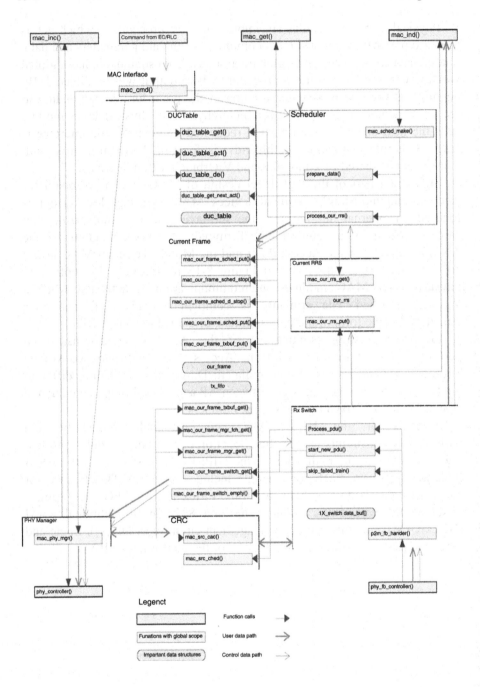

Figure 8-14. Access Point MAC layer ANSI C model - major functions and data structures

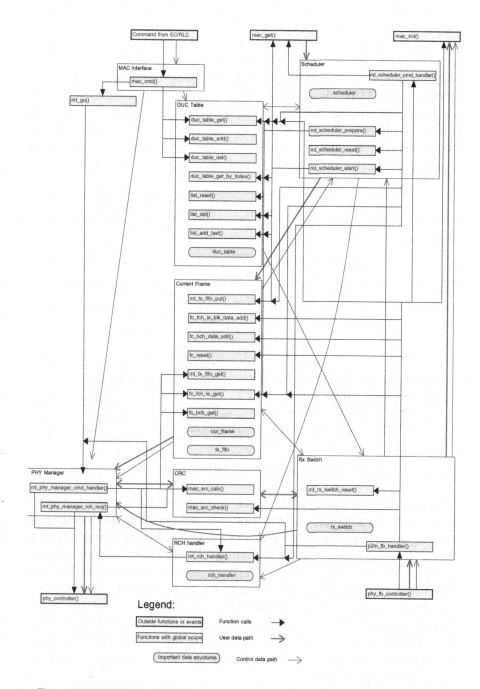

Figure 8-15. Mobile Terminal MAC layer ANSI C model - major functions and data structures

The architecture of the ARM Integrator instance that has been selected for the realization of the HIPERLAN/2 system is shown in Figure 8-16. The first core module (ARM7TDMI processor) acts as protocol processor realizing the major part of the HIPERLAN/2 DLC functionality. The second core module (ARM7TDMI processor) realizes the lower part of the HIPERLAN/2 MAC functionality and also controls the operation of the baseband block. The first logic module (Xilinx FPGA) realizes the frequency and data domain parts of the receiver. The second logic module (Xilinx FPGA) realizes the transmitter, the time domain blocks of the receiver, the interface to MAC and a slave interface to an AMBA bus.

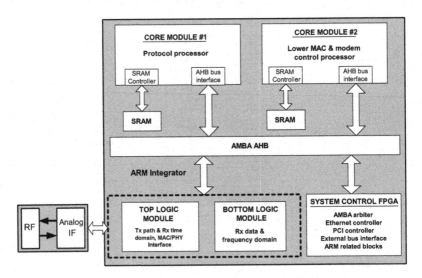

Figure 8-16. Architecture of selected ARM Integrator platform instance

It must be taken into consideration that the ARM Integrator platform is used to emulate a targeted reconfigurable System-on-Chip. Both custom and reconfigurable hardware components of the targeted reconfigurable System-on-Chip are emulated by the logic modules' FPGAs of the ARM Integrator.

System level partitioning and task assignment exploration has been performed using OCAPI-XL C++ library. Using the ANSI-C model as input, OCAPI-XL models of the HIPERLAN/2 MAC and physical layers have been developed. The block diagram of the physical layer OCAPI-XL model is shown in Figure 8-17. The block diagrams of the Access Point and Mobile Terminal MAC layer OCAPI-XL model are shown in Figure 8-18 and Figure 8-19 respectively.

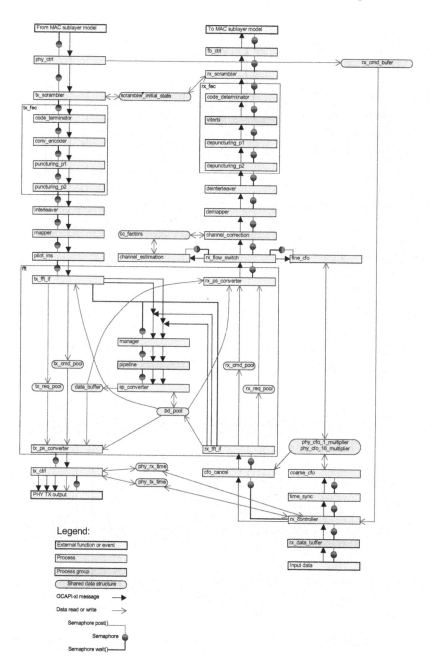

Figure 8-17. Block diagram of the physical layer OCAPI-XL model

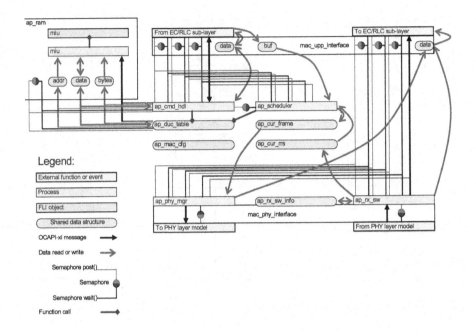

Figure 8-18. Block diagram of the Access Point MAC layer OCAPI-XL model

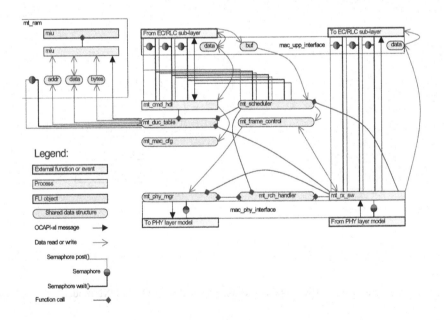

Figure 8-19. Block diagram of the Mobile Terminal MAC layer OCAPI-XL model

For the high level exploration high-level OCAPI-XL processes (procHLHW, procHLSW and procManagedSW) have been used to model the timing behavior of the HIPERLAN/2 tasks under different implementation scenarios. Using the performance estimation (in terms of execution cycles) capabilities of OCAPI-XL different mappings of HIPERLAN/2 tasks on hardware and software have been evaluated and the most promising solution has been identified.

Under the post-shipment functionality upgrading scenario, the HIPERLAN/2 system must be able to support (after upgrade) outdoor fixed wireless access operation. In that context, the tasks that are more complex, and consequently are difficult to be upgraded, are identified and assigned to reconfigurable hardware. Tasks of this kind are the receiver's channel estimation and correction block, and the receiver's decoding block (Turbo and/or Reed Solomon decoders are required in outdoor environments while Viterbi decoder is included in the HIPERLAN/2 standard).

5.   IMPLEMENTATION

The implementation phase of HIPERLAN/2 system corresponds to the detailed design and implementation design stages of the design flow described in Chapter 4 (they are also shown in Figure 8-20).

The high level OCAPI-XL model developed during high level design has been refined at a first step. The refinement included the change of processes' types from high level to low level (procOCAPI1 and procANSIC). This allowed a cycle accurate simulation of the complete system functionality and confirmation that timing constraints are met.

For the tasks assigned to instruction set processors, C code has been developed and mapped on the ARM7TDMI processors of the core modules. The tools used for the software development process include:

- *Code generation tool.* The ARM, THUMB C and Embedded C++ compilers.
- *Integrated Development Environment* Code Warrior IDE.
- *ARM Extended Debugger* Debugging environment for processor cores. It provides interface to the ARMulator and can be used to debug code on an ARM Evaluation Board.
- *Instruction Set Simulator (ARMulator)* Simulates a target system in software, allowing software development when a hardware target is not available.

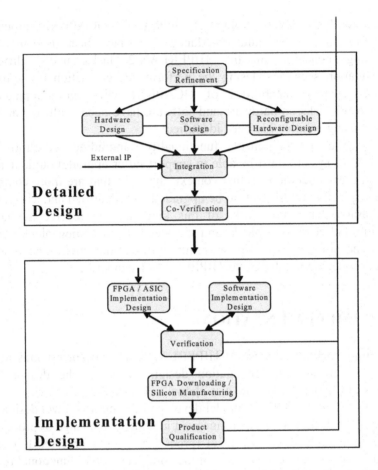

Figure 8-20. Detailed and implementation design parts of the proposed methodology

Execution times for basic tasks of HIPERLAN/2 DLC/MAC are presented in Table 8-3. The results have been obtained with an operation frequency of 50 MHz (cycle 20 ns). The code and the data for the tasks are stored in SDRAM memory.

The detailed architecture of the functionality realized by the logic modules of the platform is shown in Figure 8-21. A typical FPGA flow has been adopted for realization of the tasks assigned on the platform's logic modules (mainly base band part of HIPERLAN/2). The tools used include:

- Modelsim for simulation
- Leonardo Spectrum for synthesis
- Xilinx ISE tools for back end design

Table 8-3. Execution times for basic tasks of HIPERLAN/2 DLC/MAC layer (where AP: Access Point, MT: Mobile Terminal, CL: Convergence Layer, Tx: Transmitter, Rx: Receiver)

MT-BCH/FCH Decoder Modem Ctrl MAC Layer tasks	Execution time	DLC tasks	Execution time
Initialization Phase (Reset & Config @ slot commands)	1.20 µs	AP-Scheduler	0.2 ms
Synchronisation Phase (BCH_SRCH, Rx_FCH with rpt = 1, Rx_ACH)	2.65 µs	AP/MT-TxCL	0.6 ms
BCH decoding and BCH CRC checking	5.25 µs	AP/MT-TxBuilder (full frame)	0.7 ms
Decoding of a single IE (UL)	3.23 µs	AP/MT-TxBuilder Copy using DMA (580 bytes – word transfer)	15 µs
Decoding of 3 IEs (2 ULs, 1 DL) including CRC checking & Puncturing	15 µs	AP/MT-Rx Decoder	0.4 ms
		AP/MT-RxCL	0.7 ms

The total utilization of the bottom logic module (FPGA) is 85%. The total utilization of the top logic module is 89%. The utilization per resource type for the bottom and the top logic modules is presented in Table 8-4.

In order to fully realize the 5GHz wireless LAN access point and mobile terminal components the base band modem's functionality is followed by an IF (20 MHz to 880MHz) and an RF (880MHz to 5GHz) stage. The analog-to-digital and digital-to-analog conversion (National Semiconductors LMX5301 and LMX5306), for communicating with the IF analog front ends of the receiver and the transmitter respectively, is implemented on a separate board which seats on a dedicated connector for external communications on the "top" of the stack of logic modules. Also the communication with the PCI or Ethernet interface is done through that port.

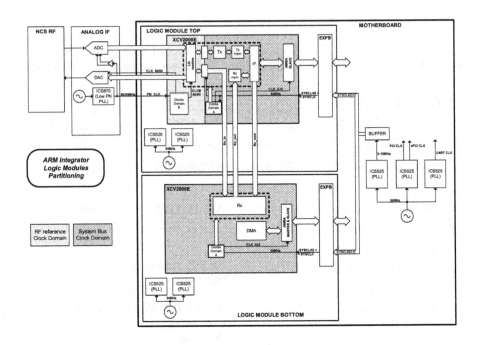

Figure 8-21. Detailed architecture of the logic modules

Table 8-4. Utilization per resource type for the two logic modules' FPGAs

	Used		Utilization	
Resource	BOTTOM Logic Module	TOP Logic Module	BOTTOM Logic Module	TOP Logic Module
I/Os	93	312	18.16	60.93
Function Generators	14923	16527	38.86	43.04
CLB Slices	12164	11252	63.35	58.60
D FFs or Latches	6368	8544	15.60	20.94

Figure 8-22 shows a photograph illustrating the ARM Integrator platform along with the IF, RF boards and the antenna, which are needed for the implementation of the access point and mobile terminal 5GHz wireless LAN components. By using two times the illustrated system (one operating as access point and a second operating as mobile terminal), a 5GHz wireless system is demonstrated. The location of each component is also indicated in Figure 8-22.

The performance results presented above from the realization of the HIPERLAN/2 system on the ARM Integrator platform are expected to improve in a reconfigurable SoC implementation. This is due to the overheads introduced by the ARM Integrator platform architecture (FIFOs of the bus interface, SDRAM controller etc.), and also due to the lack of a local bus for the communication between the base band modem and the lower MAC processor (which also controls the modem). The ARM processor running the lower MAC communicates with the base band modem and the protocol ARM through the AMBA bus, sharing the bus bandwidth with those units. This a major bottleneck with respect to real time performance of the targeted system. A System-on-Chip implementation is expected to overcome this restricted bus bandwidth because of the use of local busses as well.

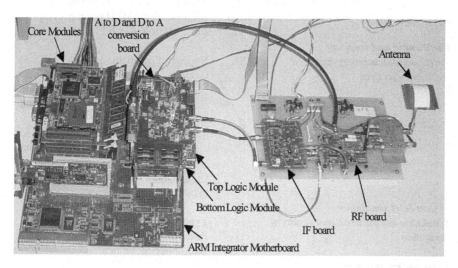

Figure 8-22. ARM Integrator platform along with the IF, RF boards and the antenna

6.   DATA FROM MEASUREMENTS AND OBSERVATIONS OF DESIGN METHODOLOGY AND TARGET SYSTEM

Data related to design methodology metrics gathered from the HIPERLAN/2 design case are presented in Tables 8-5 to 8-8.

Table 8-5. Data for application characteristics

Design flow phase/activity metrics	Data
Specification	
Size of application specification	20000 lines of ANSI-C code
Performance requirements	Throughput 54 Mbit/s
Reconfiguration objective	Flexibility towards post-fabrication upgrading

Table 8-6. Data for system-level design

Design flow phase/activity metrics	Data
Analysis *Analysis technique used to produce data (from application) for design space exploration, and main results*	Profiling (physical layer)/standard specification analysis (MAC/DLC) Number of operations (physical layer) / execution time constraint per task (MAC/DLC)
Architecture template *Main computing, storage and communication elements; reconfigurable element in more detail*	ARM Integrator platform with AMBA AHB Bus, 2 ARM7TDMI processors and 2 Xilinx Virtex E FPGAs
Partitioning and mapping *Partitions and mappings explored* *Decision objectives*	Number of partitions: 2 Contents: 1. DLC/Higher MAC on SW, time critical MAC on HW, Baseband on HW 2. DLC/Higher MAC on SW, time critical MAC on SW, Baseband on HW Criteria used: Number of execution cycles
System-level simulation *Simulation models used, and results of simulation*	Types of models and test benches: Functional Data collected: bit rate
Impact of reconfiguration *Benefits of methodology extensions to handle reconfiguration*	Unified representation of tasks that will be mapped on HW, SW, reconfigurable Hw and capability for evaluation of different assignment/partitioning options.

Table 8-7. Data for detailed design

Design flow phase/activity metrics	Data
Specification refinement *Contents of main refinements* *Rules, criteria and data used*	Complete functionality (HIPERLAN/2 MAC and baseband) OCAPI-XL
Reconfigurable part *Transformation to synthesizable description*	OCAPI-XL VHDL code generation (through OCAPI-XL refinement)/manual VHDL code development

continued

Design flow phase/activity metrics	Data
Integration	
Integration of transformed synthesizable modules, and with related software	250 VHDL files 262 C++ classes (9 top level) for the access point 258 C++ classes (8 top level) for the mobile terminal
Co-simulation, synthesis, emulation	
Technique and its characteristics used for validation of detailed design	OCAPI-XL simulation
Impact of reconfiguration	
Benefits of proposed methodology extensions	Detailed refinement path down to implementable code (HDL/ANSI C)

Table 8-8. Data for implementation design

Design flow phase/activity metrics	Data
Reconfigurable part	
Implementation design steps	Synthesis (Leonardo Spectrum) Post synthesis (Xilinx ISE) Two configuration files of 1.2 Mbytes each
Verification	
Technique and its characteristics used for verification of implementation design	On-board verification
Functional testing on Platform	
Platform, test bench and monitoring / measurement configuration	Functional performance tests (transmitted signal quality)

Data related to HIPERLAN/2 implementation metrics are presented in Table 8-9.

Table 8-9. WLAN baseband data for target system

	Metrics (target system)	Data
Performance	***Execution time***	
	Time required to execute a function or a set of functions	2 ms / frame
	Throughput	
	Rate at which samples, data, frames etc. can be processed	54 Mbit/s (worst case)
	Latency	
	Time from availability of input data until respective output is delivered	N/A
	Frequency	
	Frequencies used/achieved in the demonstrator platform	ARM7TDMI clock: 50 MHz FPGA clocks: 40 MHz, 80 MHz

continued

Metrics (target system)	Data

Area	**Size of logic and software** *Amount of logic used by reconfigurable resource, and code size of related software*	Top logic module FPGA: 312 I/Os, 16527 Function generators, 11252 CLBs, 8544 D flip flops. Top logic module FPGA: 93 I/Os, 14923 Function generators, 12164 CLBs, 6368 D flip flops. ARM7TDMI protocol processor: 1.4 Mbytes ARM7TDMI low level MAC and modem control: 50 Kbytes
Configuration	**Time** *Time required for downloading configuration onto reconfigurable resource*	1.8 seconds
	Configuration data *Amount of data required for configuration*	1.2 Mbytes for each of the two FPGAs
	Frequency/bandwidth *Frequency/bandwidth at which configuration data items can be downloaded*	680 Kbytes/second

7.   ANALYSIS OF DESIGN METHODOLOGY AND IMPLEMENTATION RESULTS

From a methodology perspective the HIPERLAN/2 system development demonstrated the benefits of adopting a seamless path from specification to implementation related stages during system development. System partitioning and task assignment alternatives (hardware, software, reconfigurable hardware) can be evaluated early in the design flow based on high-level performance estimates (produced through simulation). In this way designers may focus on the most promising alternatives in the following more detailed design stages. Furthermore the selection of architecture alternatives in a systematic way ensures that time consuming iterations from lower level design stages to architecture specifications. Both these facts reduce design time significantly.

An important drawback of the employed system level exploration methodology is that certain features of the selected implementation platform are hard to be modeled at a high level (this is especially true for board level designs). In the case of HIPERLAN/2 development the AMBA bus around which the whole system is integrated is very difficult to be included in a high level model (OCAPI-XL) with accurate performance estimates. This creates a level of uncertainty on the high level decisions made based on the high level performance estimates.

From a system point of view the HIPERLAN/2 development has proven that the exploitation of reconfigurable hardware even under a static reconfiguration scenario may lead to important benefits. The major benefit is the capability for cost efficient post-shipment functionality upgrading. In the conventional case upgrading could only happen for the tasks that are mapped to software processors in the original version. In such a case significant processing power in the form of software processors must be allocated in the implementation platform. If complex tasks that require significant parallelism (better suited for custom or reconfigurable hardware implementation) in order to meet timing constraints need to be upgraded in the future these should be realized as software from the original version already. This fact may lead to a cost inefficient realization in terms of the number and processing power (cost) of the software processors of the implementation platform. The presence of reconfigurable hardware enabling a spatial "ASIC like" computation style allows for more efficient upgrading of computationally complex tasks and thus to more cost efficient implementation platform design. The "business" advantages of the functionality upgrading scenario remain as in the software case: *short time to market* and *extended product life cycles* through upgrading derivatives.

With the described post shipment functionality upgrading scenario in mind a disadvantage is related to the design time of the tasks that will be mapped on reconfigurable hardware (to allow efficient upgrading) as compared to a software implementation (which would also allow upgrading on a software processor). This is justified by the fact that the user's programming model of most commercially existing reconfigurable devices (FPGAs) is based on HDLs (VHDL or Verilog). HDL coding of a given task takes longer than C/C++ coding of the same task. Furthermore verification and debugging but also mapping (synthesis, place and route) of HDL models take longer than the corresponding software tasks (instruction level simulation, compilation). This fact may increase the system development time depending on the amount of targeted product's upgradability.

REFERENCES

1. ETSI (2000) Broadband Radio Access Networks (BRAN); HIPERLAN type 2; Physical (PHY) layer, v 1.2.1
2. ETSI (2000) TS 101 475: Broadband Radio Access Networks (BRAN); HIPERLAN Type 2; Physical (PHY) layer
3. ETSI (2000) TS 101 761-1: Broadband Radio Access Networks (BRAN); HIPERLAN Type 2; Data Link Control (DLC) Layer; Part 1: Basic Data Transport Functions

Chapter 9

WCDMA DETECTOR

Yang Qu[1], Marko Pettissalo[2] and Kari Tiensyrjä[1]
[1] VTT Electronics, P.O.Box 1100, FIN-90571 Oulu, Finland
[2] Nokia Technology Platforms, P.O.Box 50, FIN-90571 Oulu, Finland

Abstract: The SystemC based approach and extensions are applied in the WCDMA design case in order to validate it at the system level, and to get experiences on the detailed and implementation design of reconfigurability on the selected Virtex II Pro demonstrator platform. The WCDMA detector case represents a reconfiguration scenario of applying partial dynamic reconfiguration in a mobile terminal.

Key words: Configuration overhead; design space exploration; dynamic reconfiguration; estimation; mapping; partitioning; reconfigurable; reconfigurability; SystemC; system-on-chip.

1.   INTRODUCTION

The purpose of the WCDMA detector case study is to experiment and validate the SystemC based approach for reconfigurability design. The starting point of the case study is a new adaptive channel equalizer for the WCDMA downlink [1], and the C models, test vectors and documentation of related algorithms.

The main emphasis of the experimentation is at the system-level design:
- Analysis and functional decomposition.
- High-level software and hardware estimation.
- SystemC modelling of algorithms and architecture.
- System partitioning, mapping and performance simulation.

In order to validate interfacing to the detailed and implementation design phases, the case study continues with:

N.S. Voros and K. Masselos (eds.), System Level Design of Reconfigurable Systems-on-Chips, 209-231.
© 2005 Springer. Printed in the Netherlands.

- Refinement.
- Detailed design of parts selected for run-time reconfiguration.
- Implementation of selected parts on the Virtex II Pro based platform with run-time reconfigurability.
- Measurement and analysis of properties, e.g. performance, area and configuration overhead.

The next sections describe the system, implementation platform, high-level design and implementation of the WCDMA detector case study.

2. SYSTEM DESCRIPTION

The whole WCDMA receiver base-band system is depicted in Figure 9-1, and it contains an RF and pulse shaping module, a searcher module, a detector, a de-interleaver and a channel decoder. The RF and pulse-shaping module is converting and filtering the RF signal to the base band. The searcher module performs the channel impulse response measurement and an initial synchronization. After acquiring the code phase, the searcher delivers the frame and slot synchronization signals to the detector (grey area in Figure 9.1). The detector functionality is explained in detail later. After detection, the de-interleaver carries out the de-interleaving, de-multiplexing and rate de-matching tasks [2]. Finally the channel decoder functionality is performed.

The case study focuses on the detector portion of the receiver. The case study covers only a limited feature set of the full receiver. The detector case uses 384kbit/s user data rate without handover.

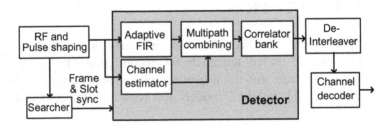

Figure 9-1. WCDMA receiver base-band system.

2.1 Detector Architecture

The detector (grey area in Figure 9-1) contains an adaptive filter, a channel estimator, a multi-path combiner and a correlator bank. The adaptive filter is performing the signal whitening and part of the matched filtering implemented traditionally with the RAKE receiver. The channel estimator module calculates the phase references. In the combiner part, the different multi-path chip samples are phase rotated according to the reference samples and combined. Finally the received signal is de-spread in the correlator bank.

When compared to traditional RAKE based receiver concepts, this WCDMA detector achieves 1 − 4 dB better performance in vehicular and pedestrian channels. The detector provides thus performance benefits in more challenging channel conditions. As the traditional RAKE concepts contain several correlators for tracking the multi-path components, this detector contains a single channel equalizer for performing multi-path correction. This results in improved scalability, since increasing multipaths or data rates would mean increasing amount of early/on-time/late correlators in the traditional RAKE based concepts.

2.1.1 Adaptive Filter

Regardless of the data rates or channel configurations required by the specification, the adaptive filter block is unchanged as it simply processes chip samples before the de-spreading takes place. Extendibility aspects are also not a problem as no changes are required to support other demands.

The adaptive filter is implemented by using basic FIR filtering structures with a delay line and taps for extracting complex sample values to be multiplied with the tap coefficients. The implementation is fully parallel, so the number of multiplier units for coefficient multiplication in both I and Q branches and the units needed for calculating new coefficients equal the number of taps in the filter.

2.1.2 Channel Estimator

The function of the estimator is to de-spread the CPICH (Common Pilot Channel) chips on different multi-paths with correctly timed spreading and channelization codes. Then the received and de-spread CPICH symbols are multiplied with the complex conjugates of the CPICH pilot pattern. The output is channel estimates for different multi-paths, which are used in the combiner to rotate received chips in different multi-paths before combining, in order to match their phases and amplitudes.

The channel estimator receives timing information from the searcher block. This includes the delay information about multi-paths at a specified delay spread. The channel can therefore be thought as a FIR filter with a number of taps and with most taps zero-valued. The task of the channel estimator is to find the tap values for those taps that the searcher determines to be non-zero.

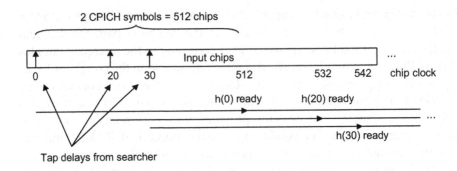

Figure 9-2. Channel estimation example.

In an example of Figure 9-2, the searcher has found three non-zero taps for a certain base station (delays 0, 20 and 30). The multi-path combining of the received chips c is then done in the combiner by multiplying received chip samples with the complex conjugates of the channel estimates h, after all multi-path estimates are ready.

In the DPCH-channels (Dedicated Physical Channel), the pilot symbols can be found from the end of the slot, while the TPC (Transmit Power Control) symbols are located in the beginning half of the slot as depicted in Figure 9.3.

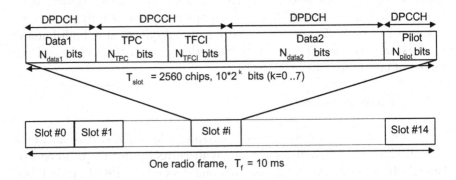

Figure 9-3. Downlink DPCH L1 frame structure.

The CPICH channel estimate over one slot is formed by integrating over the number of symbols and then it is scaled. It is used for actual phase correction of the received chips. The CPICH estimates are used as channel references for every data channel.

2.1.3   Combiner

As the base station transmits the pilot symbols through the channel, the terminal receives the directly propagated symbols and the delayed multipaths. As the pilot symbols are known beforehand, the channel tap coefficients for each multi-path can be calculated.

The different multi-path chip samples are first phase compensated according to the channel tap estimates. This is done by multiplying the chip sample with the complex conjugate of the corresponding multi-path channel tap coefficient. Finally all the phase compensated chip samples are added together to form an equalized chip sample.

2.1.4   Correlator

The function of the correlator bank is to create de-spread symbols from the output of the multipath combiner. Combined chips are de-spread by spreading code, which is formed from scrambling and channelization codes. After de-spreading, chips are integrated over the symbol period in an integrator and the result is scaled.

3.   IMPLEMENTATION PLATFORM DESCRIPTION

The implementation platform is Memec Design's Virtex-II Pro FF1152 P20 Development Kit. The board architecture is shown in Figure 9-4.

The onboard FPGA, Xilinx Virtex2P XC2VP20, contains two embedded PowerPC 405 processors, 18560 LUTs, and 88 BlockRAMs.

The system board includes two independent memory blocks of 8 M x 32 SDRAM, five clock sources, two RS-232 ports, a high-speed 16-bit LVDS interface supporting SPI-4.2, two iSFP Gigabit Ethernet optical interfaces, a 10/100 Ethernet PHY, an IDE connector, and additional user support circuitry to develop a complete system.

The System ACE interface on the Virtex-II Pro system board provides the flexibility to store multiple FPGA configuration options in removable Compact Flash cards.

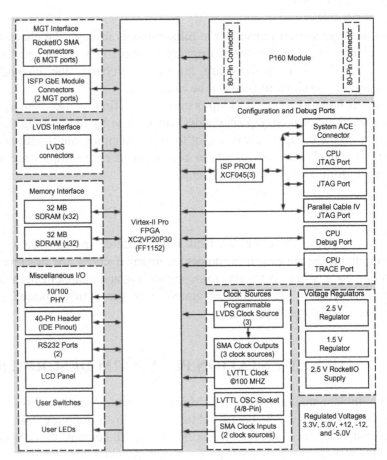

Figure 9-4. Virtex II Pro FF1152 development board architecture (Memec Design).

4.   HIGH LEVEL DESIGN

The proposed high-level design methodology and reconfigurability extensions to SystemC based approach are applied in the WCDMA detector case study. The system-level design part of the design flow is depicted in Figure 9-5.

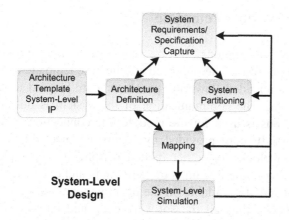

Figure 9-5. System-level design part of the proposed methodology.

Given the C codes of the algorithms of the WCDMA detector as the starting point, the following system-level design steps were performed:

- Analysis/functional decomposition of specification.
- High-level estimation.
- SystemC modelling of algorithms and architecture.
- System partitioning, mapping and performance simulation.

4.1 System Specification

System specification is given as a combination of documents and C codes of the functions of the system. Based on these a complete C-based description of the system functionality is created. The structure of the WCDMA detector C codes is depicted in Figure 9-6.

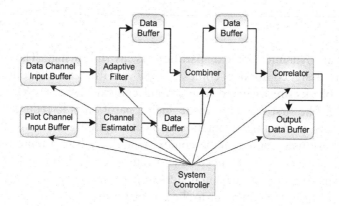

Figure 9-6. Structure of WCDMA detector C codes.

4.2   Architecture Template

The architecture options to be considered are limited by the implementation platform. The proposed architecture template is described as following. The embedded PowerPC processor core is used to perform the software functions. The communication of the system is not complicated, and one system bus is used for the module interconnection. The Processor Local Bus (PLB) IP core provided by Xilinx is selected. The WCDMA detector function blocks and other related peripherals are implemented as either dynamic reconfigurable units or static units and connected to the PLB bus. The two SDRAMs available on the development board are not suitable for fast data access, thus all the software code and intermediate data are stored in the on-chip block RAMs.

4.3   Estimation and Partitioning

The estimation approach and prototype tool described in Chap. 5 are used to estimate resources required by the function blocks on a Virtex II Pro-type FPGA. The results are presented in Table 9-1, where LUT stands for look-up tables and register refers to word-width storages. The multiplier refers to the hardwired 18x18 bits multipliers embedded in the Virtex II Pro FPGA.

Table 9-1. Estimates of FPGA-type resources required by function blocks.

Functions	LUT	Multiplier	Register
Adaptive filter	1078	8	91
Channel estimator	1387	0	84
Combiner	463	4	32
Correlator	287	0	17
Total	3215	12	224

The final dynamic context partition is as following. The channel estimator is assigned to one context (1387 LUTs), and the other three processing blocks are assigned to a second context (1078 + 463 + 287 = 1828 LUTs). The first reason to make such a partition is to reduce the total amount of area in the implementation. However, putting the correlator and the channel estimator into a single context would result in a more balanced partition. This option is dropped due to the consideration of the data transfers. The amounts of data transferred from the adaptive filter to the combiner and from the combiner to the correlator are much higher than that of other paths, so putting the adaptive filter, the combiner and the correlator into one context will also reduce the interface complexity.

4.4 Mapping and System-Level Simulation

A fixed system is created first, which has two purposes in the design. The first one is to use its simulation results as reference data, so the data collected from the reconfigurable system can be evaluated. The second purpose is to use it as the input to the DRCF transformer to generate the DRCF component.

In the fixed system, all of the four detector functions are mapped onto separate hardware accelerators and the scheduling task is mapped onto a software task that runs on the PowerPC processor core. In the scheduling, pipelined processing is used to increase the performance. A small system bus is modelled to connect these processing units. The channel data for simulation is recorded in text files, and the processor drives a slave I/O module to read the data. The SystemC models are described at the transaction level, in which the workload is derived based on the estimation results but with manual adjustment. Part of the waveforms generated from the simulation results is presented in Figure 9-7. The results show that 1.12 ms is required for decoding all 2560 chips of a slot when system is running at 100 MHz.

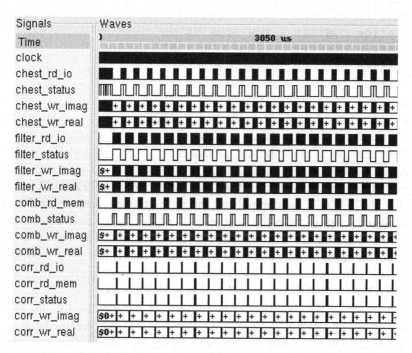

Figure 9-7. Simulation waveform of the fixed system for the detector.

The dynamically reconfigurable system is generated in a way that follows the context-partitioning decision, which is to assign the four processing blocks into two contexts. The DRCF transformer is used to replace these four static accelerator modules with a DRCF component. The code below shows the script file for the DRCF transformer.

```
// for DRCF transformer
TOP_LEVEL : top top.h
/* bus interface info */
BUS_INTERFACE: bus_mst_if bus_if.h
/* slave interface info */
SLAVE_INTERFACE: bus_slv_if bus_if.h
/* modules info */
MODULES: adp_filter adp_filter.h, comb comb.h,
         corr corr.h
MODULES: chest chest.h
// DRCF_configure.h
#define DRCF_PRI 5
#define DRCF_SIZE 100
#define CONTEXT_1_LOW_ADDR 0
#define CONTEXT_1_LENGTH 76754
#define CONTEXT_1_SIZE 1387
#define CONTEXT_2_LOW_ADDR 76755
#define CONTEXT_2_LENGTH 101158
#define CONTEXT_2_SIZE 1828
#define CONFIG_BITWIDTH 16
#define DRCF_MEM_LOW_ADDR 0
#define DRCF_MEM_HIGH_ADDR 180000
#define DRCF_MEM_READ_LATENCY 1
#define DRCF_MEM_WRITE_LATENCY 1
#define RECONFIG_CLK_DIV 3
```

In the code, the *TOP_LEVEL*, *BUS_INTERFACE*, and *SLAVE_INTERFACE* are compulsory macros so the transformer can find the required information from the source code of the initial architecture. Each macro of *MODULES* will initialize a new context. The names of the SystemC modules that are assigned to the new context are given next to the *MODULES*. In this case, the *adp_filter*, *comb* and *corr* are given next to the first *MODULES*, and the *chest* to the second *MODULES*.

The group of *#define* are the definitions of the macros for the DRCF component. The definition *CONTEXT_N_LENGTH* represents the size of the bit stream of the context *N*. In this case, we assume that the size is proportional to the resource utilization, which is the number of LUTs required. The total available LUTs and size of full bit stream are taken from the Virtex2P XC2VP20 datasheet. The final model is depicted in Figure 9-8.

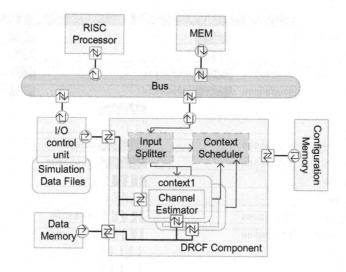

Figure 9-8. SystemC modelling of the WCDMA detector module.

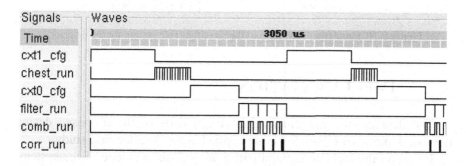

Figure 9-9. Simulation waveform generated by the DRCF component.

The performance simulation is performed after the creation of the reconfiguration system. The system requires 2 reconfiguration requests per slot. When the configuration clock is running at 33 MHz, the reconfiguration latency is 2.73 ms for 16 bits configuration bit-width. The solution is capable of decoding 3 slots in a frame. The configuration status generated by the DRCF component in the simulation is shown in Figure 9-9. The simulation results for the reconfigurable system are shown in Figure 9-10.

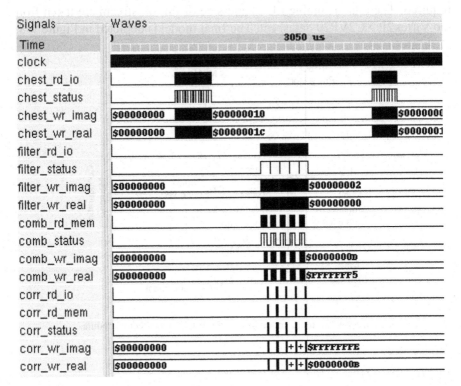

Figure 9-10. Simulation waveform of the reconfigurable system for the detector.

5.   IMPLEMENTATION

The detailed and implementation design parts of the proposed design flow as, depicted in Figure 9-11, are applied in the detector case study.

Based on the architecture and partitioning decisions done in the system-level design, the following design steps are performed:

- Refinement and detailed design of parts selected for reconfiguration.
- Implementation of selected parts on the Virtex II Pro platform.
- Measurement and analysis of properties, e.g. performance (cycle counts), area (logic blocks) and configuration overhead (time).

The design flows and tools provided by Xilinx are used. The EDK provides tools for the Virtex II Pro platform integration and software development on PowerPC processor. The ISE provides tools for the FPGA design. Bit streams of the dynamic contexts are generated using the modular design flow [3], and the reconfiguration bit streams are downloaded to the Virtex II Pro FPGA by using the SystemACE module.

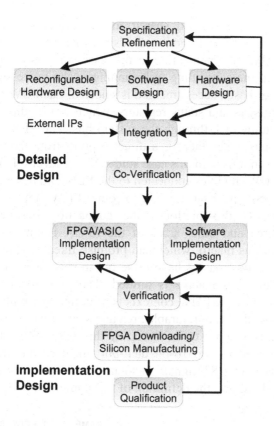

Figure 9-11. Detailed and implementation design parts of the proposed methodology.

5.1   Specification Refinement

In this step, the major constraints come from the limitations of the Virtex-II Pro platform and low-level implementation tools. The inputs are the C functions of the system and the SystemC model of the system, and the refinement focuses on the interface refinement and the DRCF component refinement.

5.1.1   Interface Refinement

The task is divided into traditional SW/HW interface refinement and common interface refinement of RHWs, which refers to the two dynamic contexts presented in the previous section. The common interfaces of RHW modules refer to the common signals that connect the RHWs to the rest of the system.

Direct register-based access, shared memory access and interrupted access are the main communication methods between SW and HW. Because the primary SW/HW communication in this case study is the triggering of HW modules from the scheduling task running in the processor, the direct SW-to-HW register-based access method is selected. Each HW module has a few command registers and status registers, which can be directly accessed by the SW scheduling task through the system bus. The PLB is selected as the main system bus, since there is a direct support from the EDK tool. A second level bus is not used in this case study. Xilinx level-0 I/O SW drivers are used to drive HW modules connected to the PLB, and bus adapters based on the PLB IPIF SSP1 package are used to connect HW modules to the PLB.

In the Virtex II Pro platform, the connection between a partial reconfiguration area and other areas of the FPGA is implemented via a special unit called bus macro, which can promise fixed routing of 4 signals each. However, the bus macros have to be put in specific locations and the number of available bus macros is limited. The purpose of the common interface refinement of the RHW contexts is to define the minimum number of signals that cross the reconfigurable area and static area in order to reduce the number of bus macros to be used. In this case study, the common interfaces are reduced to two 16-bit dual-port memory interfaces and one PLB IPIF interface with 82 signals in total, which corresponds to 21 bus macros. Figure 9-12 shows the definition of this common interface.

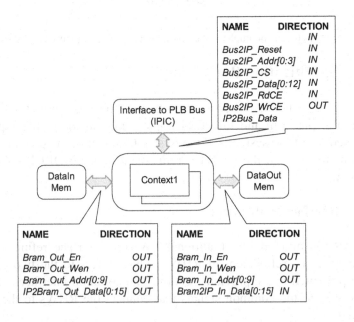

NAME	DIRECTION
	IN
Bus2IP_Reset	IN
Bus2IP_Addr[0:3]	IN
Bus2IP_CS	IN
Bus2IP_Data[0:12]	IN
Bus2IP_RdCE	IN
Bus2IP_WrCE	OUT
IP2Bus_Data	

Interface to PLB Bus (IPIC)

DataIn Mem

Context1

DataOut Mem

NAME	DIRECTION
Bram_Out_En	OUT
Bram_Out_Wen	OUT
Bram_Out_Addr[0:9]	OUT
IP2Bram_Out_Data[0:15]	OUT

NAME	DIRECTION
Bram_In_En	OUT
Bram_In_Wen	OUT
Bram_In_Addr[0:9]	OUT
Bram2IP_In_Data[0:15]	IN

Figure 9-12. The common interface for RHW contexts.

5.1.2 Configuration Refinement

In this step, the task is to decide when and how to trigger the reconfiguration. The behaviour of the DRCF component at the system level is to automatically generate reconfiguration overhead when needed. In the low-level implementation, the triggering of the reconfiguration is embedded in the SW code. The place where to trigger the reconfiguration is extracted by analysing the SystemC simulation results, which record the simulation time and conditions when new reconfigurations are triggered. In this case, the function calls both to the adaptive filter and to the channel estimator will trigger reconfiguration.

The ICAP and the SystemACE Compact Flash (CF) solution are the two options to allow the embedded processors to manage the reconfiguration of the system at run time. The ICAP is an in-chip solution, which is able to reconfigure an individual CLB or a frame. The SystemACE module, which is available in the development board, provides a space-efficient, pre-engineered, high-density configuration solution. The SystemACE CF solution supports the use of CF cards up to 256 Mb, which makes it easier to store and manage the reconfiguration bit stream. In the case study, the SystemACE CF solution is selected.

5.2 Design of Modules

5.2.1 Implementation Architecture

Figure 9-13 shows the implementation architecture of the detector module on the Memec Virtex-II Pro platform. The design of the system is divided into 3 separate tasks: the design of static module, the design of dynamic module and the design of SW.

The static module contains all the storage units and computation units that are not supposed to be reconfigured when the system is running. In the case study, the static module contains the PPC hard core, the data BRAMs, the instruction BRAMs, SystemACE controllers, PLB, reset module, memory controller module and other peripheral modules to connect to the outside world. The EDK design suite provides most of the IP modules required in the static module implementation. The others are manually coded.

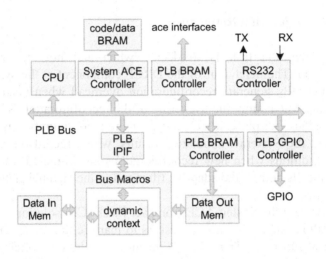

Figure 9-13. Implementation architecture on Virtex II Pro with RHW approach.

5.2.2 SW Design

The software design phase mainly includes the extraction of the scheduling task from the original SystemC code. The EDK SW implementation tool is C-oriented implementation platform, so majority of the function does not need to change. The main task involved is to replace the *read()* and *write()* interface method calls in the SystemC code with *XIo_In16()* and *XIo_Out16()* IO functions, which are available as level-0 I/O SW drivers.

The run-time reconfiguration is triggered and managed by SW code. The SystemACE module is physically attached to the development board and controlled by the SystemACE controller IP, which is available from Xilinx. The SW routine to trigger the reconfiguration process is also implemented using the level-0 I/O SW drivers, and the routine is inserted into system SW code with the guide of the configuration refinement results.

5.2.3 RHW Module Design

The process of RHW module design is the manual translation from C/C++ application code to RTL models. First, each of the four detector functions is implemented as a single block. Then, a common context wrapper, which contains the common interface signals as shown in Figure 9-12 above, is used to wrap the channel estimator as one module and the other three as another module. In the second module, multiplexers are used to solve the conflicts of the output signals of the three blocks. In the implementation phase, which will be presented in later sections, each module is implemented as a partial bit stream.

The design of the combiner and the correlator are the same as they are in a fixed system. The adaptive filter and the channel estimator are the processes with memory. Their outputs depend on both the input values and the internal states of the process. In order to maintain the internal states of the two modules, separate memory blocks are used to hold the internal states between reconfiguration and proper changes are made in the VHDL RTL code to perform the save and load of the internal state information.

5.3 Integration and Co-Verification

The EDK design suite is used to create the simulation files for the complete system. However, the tool set does not provide the support to integrate the two dynamic contexts and the static part of the system into a single simulation environment. Two systems for simulation are created, and each one is simulated individually. The first one is the integration of the static part and the context-1 module, and the second integration is for the static part and the context-2 module. The output of the system containing the context-1 module (channel estimator) is manually fed to the system containing the context-2 module. The reconfiguration delay is estimated according to the systemACE datasheet. ModelSim is used as the simulation platform. VHDL RTL code of the context modules and other peripherals are directly fed into the ModelSim tool. The SW code is compiled and converted to Block RAM data. The PPC core is simulated using SmartModel, which can be directly linked to the ModelSim using the SWIFT interface to perform SW/HW co-simulation.

5.4 SW and RHW Implementation Design

The implementation processes of the WCDMA detector system are described in the following sections.

5.4.1 Software Implementation Design

Xilinx provides the GNU toolkit for the software implementation process. Both the GCC compiler and GDB debugger are available in the EDK design suite. Inputs are the C code and EDK-generated header files. The output is a compiled binary file in ELF format, which can be directly downloaded to the FPGA. The same binary file is also used in the SW/HW co-simulation step.

The size of the SW code is shown in Table 9-2. When level-0 SW driver libraries are included, the total size is 181K bytes.

Table 9-2. SW implementation results.

	Text	Data	Un-initialized data	Total
Size (byte)	43588	4152	8248	55988

5.4.2   RHW Implementation Design

The Synplify Pro is used to synthesize the VHDL RTL models of the WCDMA detector functions. The results are shown in Table 9-3.

The module-based partial reconfiguration design flow [3] is used to implement the two contexts in the Virtex II Pro. There are three parts to be implemented, one static part and two run-time reconfigurable contexts that are overlapped with each other in the same area. In the implementation, 920 LUTs and 4 Block RAMs are required for the context containing the channel estimator, and 1254 LUTs, 6 Block RAMs and 12 Block Multipliers are required for the other context. The static part requires 1199 LUTs and 25 Block RAMs. There are 21 bus macros used to connect the static part and one dynamic context. The decoding time of one slot of data is 9.66 ms including the reconfiguration latency.

Table 9-3. HW synthesis results.

	LUT	18x18 Multiplier	Register bits
Adaptive filter	553	8	1457
Channel estimator	920	0	2078
Combiner	364	4	346
Correlator	239	0	92

The case study uses 36 external IO pins, and 33 of them are located in the right side of the FPGA. Because an IO pad is also part of reconfigurable resources, in order to maintain the connections to the IO pads are fixed during reconfiguration, the static part is assigned to the right side of the FPGA (SLICE_X44Y111:SLICE_X91Y0) and the contexts are assigned to the left side of the FPGA (SLICE_X0Y111:SLICE_X43Y0). The 21 bus macros are inserted in between the static part and the dynamic part. The one IO pad that is located in the left side of the FPGA is routed to the right side via a bus macro. The size of the partial bit streams generated for the context-1 and the context-2 are 278k bytes and 280k bytes respectively.

A routed design after module assembly is shown in Figure 9-14. The assembled design is the integration of the context1, in the left side, and the static part, in the right side. The bus macros for signal connection are shown in the middle block.

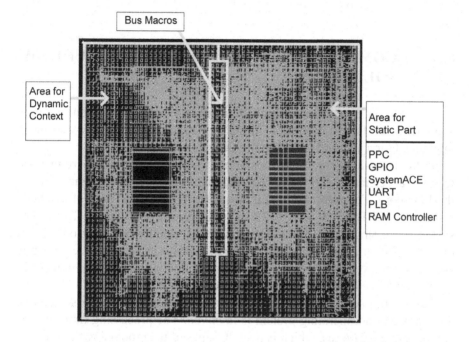

Figure 9-14. Routed Design of the assembly of the context 1 and the static part.

The data collected for the whole system is given in Table 9-4. The achievable clock frequency is 101 MHz.

Table 9-4. Xilinx Virtex II Pro XC2VP20 resource utilization.

	LUT	Block RAM	18x18 Multiplier	Register (bit)	PPC hard core
Static part	1199	41	0	1422	1
Dynamic part	1534	7	12	1855	0
Total	2733	48	12	3277	1

5.5   Downloading and Execution

The iMPACT tool is used to transform the configuration files into SystemACE file format, segment the space of the CF card and perform the necessary file management. The CF card writer is used to store the transformed files in a 128 MB CF card.

In the execution, a complete system (integration of the static part and the context1) is initially downloaded to the FPGA using the iMPACT, and the partial bit streams are loaded when necessary by the SystemACE module. In this case study, the RS-232 and LCD peripherals are added to the PLB system bus for displaying messages and results.

6.   COMPARISON WITH FIXED HW AND PURE SW SOLUTIONS

In addition to the implementation of the dynamic reconfiguration approach, a fixed hardware implementation and a pure software implementation are made as reference designs.

In the fixed-hardware implementation, the processing blocks are mapped to static accelerators and the scheduling task is mapped to SW that runs on the PPC core. The resource requirements are 4632 LUTs (24% of available resources), 55 Block RAMs (62%) and 12 Block Multipliers (13%). The system is running at 100 MHz. The processing time for decoding one slot of data is 1.06 ms. Compared to the fixed reference system, the dynamic approach achieves almost 50% resource reduction in terms of the number of LUTs, but at the cost of 8 times longer processing time.

For the full software implementation, the design is done as a standalone approach and no operating system is involved. The processing time for one slot of data is 294.6 ms, which is over 30 times of the processing time in run-time reconfiguration case. This does not fulfil the real-time requirements.

7.   RESULTS ANALYSIS

Results related to design methodology and implementation are separately discussed in the following sections.

7.1   Analysis of Design Methodology Results

7.1.1   Advantages

The main advantage of the SystemC-based approach is that it can be easily embedded into a SoC design flow to allow fast design space exploration for different reconfiguration alternatives without going into implementation. The decision of the context partitioning is guided by the SystemC-based design methodology. The reconfiguration effects are modelled through parameters, whose values are annotated from the data of the target platform.

Timing/resource estimation, DRCF modelling and performance simulation are the main techniques to support fast design space exploration at the system level. Different alternatives can be compared in order to make justified decisions of allocating functions to dynamic contexts. The automatic SystemC code generation for the reconfigurable module can significantly reduce the coding effort. The transaction-level modelling enables fast performance evaluation of reconfiguration effects at the system level. Considering the design at detailed level and implementation level is time consuming, usually taking from weeks to months, the SystemC-based approach can result in remarkable improvements in the design process both from time and quality point of view.

7.1.2   Disadvantages

From the methodology point of view, the lack of study of power performance is clearly a disadvantage, since power is an important issue in design of wireless equipment. The power variation during the run-time reconfiguration process needs to be addressed.

The link from the system level to the detailed-design level involves manual transformation from C to HDL, which tends to be time-consuming and error-prone. High-level synthesis tools could be candidate approach. Although falling out of the scope of the current research, improvements to the vendor-specific design flows and tools for the detailed and implementation design could be welcome, too.

7.1.3   Summary

It is very important to have an approach that allows designers in the early phase of design to rapidly explore the differences of using different reconfiguration alternatives. The SystemC-based instantiation of the design flow introduced in Chapter 4, has been proven its applicability in practice through the successful design of the WCDMA detector case study. The use of run-time reconfigurable hardware will create a flexible system and result in shorter time-to-market when comparing with equivalent ASIC-type SoC implementation.

7.2   Analysis of Implementation Results

7.2.1   Advantages

The potential benefit of using run-time reconfiguration approach is obviously the significant reduction of reconfigurable resources. Compared to a completely fixed implementation, the reduction of LUTs can be up to 50%. Compared to a fully software implementation, the run-time reconfiguration approach is over 30 times faster.

7.2.2   Disadvantages

The commercial off-the-shelf FPGA platform caused limitations on the implementation of run-time reconfiguration. Although the selected approach used partial reconfiguration, the required configuration time affected the performance a lot in the data-flow type WCDMA detector design case. The design case was possible to be implemented by using the vendor-specific design flows and tools for the detailed and implementation design, but some manual work-around were needed.

The run-time reconfiguration implementation of the WCDMA detector resulted in severe reconfiguration latency, which however is due to the limitation of the FPGA technology. The ratio of computing to configuration time was about 1/8 in this design case. The reconfiguration latency has been revealed in the SystemC simulation using the DRCF modelling technique. The overall performance is expected to be significantly improved when advanced approaches are available, such as multi-context devices.

7.2.3   Summary

The dynamic partial reconfiguration of the WCDMA detector was designed and implemented on a commercial Virtex-II Pro FPGA-platform in order to validate the system-level extensions of the SystemC based approach. The validation results showed the system-level approach to be valid. The implementation results showed long reconfiguration latency in the detector design case, although also demonstrating possibilities for resource sharing when compared to the fixed hardware implementation and for performance improvement over pure software implementation.

8.   CONCLUSIONS

The goal of the design case is to use SystemC-based approach to study the feasibility of dynamic reconfiguration of a new detector algorithm. The design starts from C code. The SystemC-based approach and tools are used for early design space exploration. Commercial tools and manual VHDL coding are involved in the detailed-design level and in the implementation level. The dynamic partial reconfiguration design presents 40% area saving but 8 times longer processing time when compared with a fixed hardware implementation. When compared with a pure software solution, it presents over 30 timers better performance.

The commercial off-the-shelf FPGA platform caused limitations on the implementation of run-time reconfiguration. Although the selected approach used partial reconfiguration, the required configuration time affected the performance a lot in the data-flow type WCDMA detector design case. The implementation of the WCDMA detector demonstrator validated that the SystemC-based approach and associated support tools are able to support the design of reconfigurable SoCs at the system level. Timing/resource estimation, DRCF modelling and performance simulation are the main techniques to support fast design space exploration at the system level. Different alternatives can be compared in order to make justified decisions of allocating functions to dynamic contexts. Consequently, iterations from detailed and implementation design back to system-level can be avoided.

k to system-level can be avoided.

REFERENCES

1. M. J. Heikkila (2001) A novel blind adaptive algorithm for channel equalization in WCDMA downlink", In: The 12th IEEE International Symposium on Personal, Indoor and Mobile Radio Communications, 2001, Volume: 1, Pages: A-41-A-45.
2. 3GPP TS 25.201: 3rd Generation Partnership Project; Technical Specification Group Radio Access Network; Physical layer – General description (Release 6).
3. Xilinx. Inc., XAPP 290: Two Flows for Partial Reconfiguration: Module Based or Small Bit Manipulations. v1.0, May 2002.